Prevention of Pressure Sores

Prevention of Pressure Sores

Edited by

J G Webster

Department of Electrical and Computer Engineering,
University of Wisconsin-Madison

CRC Press
Taylor & Francis Group
Boca Raton London New York

CRC Press is an imprint of the
Taylor & Francis Group, an **informa** business
A TAYLOR & FRANCIS BOOK

First published 1991 by Taylor & Francis

Published 2019 by CRC Press
Taylor & Francis Group
6000 Broken Sound Parkway NW, Suite 300
Boca Raton, FL 33487-2742

First issued in paperback 2019

ISBN 13: 978-0-367-45600-9 (pbk)
ISBN 13: 978-0-7503-0099-5 (hbk)

Visit the Taylor & Francis Web site at
http://www.taylorandfrancis.com

and the CRC Press Web site at
http://www.crcpress.com

Library of Congress catalog number: 90-5395

Library of Congress Cataloging-in-Publication Data

Catalog record is available from the Library of Congress

Contents

CONTENTS

Preface

This book covers the causes and description of pressure sores, also called decubitus ulcers, bed sores, or pressure ulcers. A pressure sore is an ulceration caused by excessive pressure being applied to a tissue over an excessive duration. The book details the methods of pressure sore prevention in chairs, beds, and on the operating table. It discusses the methods of measuring pressure distribution at the tissue/support surface interface and the accuracy of these measurements.

Most books are very descriptive and have been written by nurses. The quantitative information available from rehabilitation engineers is scattered throughout journals and conference proceedings. The purpose of the book is to collect in one place the diverse knowledge of pressure sores. It critically evaluates studies by rehabilitation engineers, orthopedic surgeons, and nurses. It helps each of these groups and patients understand cause and effect for pressure sores. Everyone agrees that they are preventable, but many pressure sores still occur. This book describes how to prevent them.

Chapter 1 details the causes of pressure sores, their incidence, stages of development and the contribution of mechanical factors. Starting with physical characteristics of soft tissue, Chapter 2 shows how computer models predict the pressure distribution at various depths and how the pressure sore may first develop at depth, rather than at the surface.

The early warning signs and risk assessments in chapter 3 and the behavior to prevent pressure sores in chapter 4 should be useful to both caregivers and patients.

Chapter 5 describes the many types of seat cushions—the most-used preventive device. It also includes design and evaluation methods for seat cushions. Other support surfaces, such as hospital beds and operating table pads are described in chapter 6. The complex pressure relief methods described in chapter 7 include alternating pressure devices, and air, sand, water, and net beds.

Chapter 8 provides the engineering details of operation of the bladder pressure sensors and chapter 9 details of conventional pressure sensors used for measuring interface pressures. Chapter 10 shows the pressure distribution visualization provided by simple plastic-on-glass displays and those provided by more intricate methods on computer screens. Accuracy and inconsistencies of pressure measurement are detailed in chapter 11, as well as measurement of shear—which many feel is more important than pressure.

Wheelchair pressure reliefs are recommended for preventing pressure sores. Chapter 12 describes the relation between pressure and duration and how this is used in systems that provide alarms to the patient. Chapter 13 presents other possible preventive methods such as functional electrical stimulation, soft tissue augmentation, and drug-based prevention.

Chapter 14 describes the treatment of pressure sores, including the physiology of would healing, assessment and classification, and costs of treatment. The 400 references in chapter 15 provide an extensive source for further study. Each chapter provides a list of study questions suitable for testing comprehension in self study or in the classroom.

The book should prove invaluable for rehabilitation engineers who design and test support surfaces for preventing pressure sores. It should also prove useful for nurses and other health care providers who work with patients to prevent and treat pressure sores. Patients at risk of pressure sores should also gain an understanding about causes, prevention, and treatment to improve their own prognosis.

All contributors are from the Department of Electrical Engineering at the University of Wisconsin–Madison and I wish to thank them for their efforts in creating this book.

John G Webster
Madison, Wisconsin
November 1990

1

The Cause of Pressure Sores

Jon Pfeffer

A pressure sore is an ulceration caused by excessive pressure being applied to a tissue over an excessive duration. These ulcers are also called decubitus ulcers, bed sores, or pressure ulcers; these names implying that they are caused by lying down too long. We will use the term *pressure sore* since pressure is the main factor that causes this type of ulceration, regardless of what posture or position the person is in when a sore is acquired.

1.1 INCIDENCE OF PRESSURE SORES

1.1.1 Frequency of occurrence

Elderly people represent the single largest group likely to suffer from pressure sores (Torrance 1983). With medical care gradually improving around the world, life expectancies are increasing, which leaves a larger segment of the population susceptible to pressure sores. Barton and Barton (1981) show that hospitals with the most advanced wards had higher incidences of pressure sores compared to less developed wards, which had a higher death rate. Periodic surveys have counted the number of people who have developed a pressure sore among selected groups of people. Surveys of general hospitals indicate that from 3 to 4.5% of patients develop pressure sores during hospitalization (Manley 1978, Petersen and Bittmann 1971). Allman (1989) stated that the prevalence of pressure sores was 3 to 11% in acute care hospitals and nursing homes. In a national study in England and Wales, David *et al* (1983) found that 85% of the 885 patients reported to have pressure sores were over the age of 65.

From the surveys taken, it is desirable to know the probability of pressure sore development and the total number of pressure sores occurring in each group. Several examples of groups that are susceptible to pressure sore development include the spinal cord injured, the elderly, the incontinent, orthopedic patients, and patients in intensive care wards. A person can be a member of more than one group at a time, thus making it difficult to identify which group is truly more susceptible to pressure sore formation. Figure 1.1 lists a collection of surveys by Torrance (1983). These surveys indicate that people over the age of 70 had a high incidence of pressure sores. For example, in Woodbine's survey B, 75% of the people that developed a pressure sore were over the age of 78. In Gerson's (1975) study the average age was 49.5 years, but the average age of those with pressure sores was 67.7 years. For the interested reader, a more complete description of these surveys is found in chapter 1 of Torrance's book. It should be noted that more surveys than these have been published; however, it is difficult to compare them due to the different methods and gradings of pressure injury that were used.

Study	Population	Number with sores	Number of sores	% people with sores	% people of age 70+
Petersen and Bittmann (1971)	Total 517,000 Hospitalized (4,437)	223 (131)	318 (188)	0.43 (3.0)	15.5
Jordan and Clark (1977)	10,751	946	1,394	8.8	50
Stapleton (1978)	2,218	135	211	6.0	45.6
Ek and Bowman (1982)	1,776	71	109	4.0	73.8 years mean ave
Jordan and Nicol (1977)	999	94	____	9.4	70
Gerson (1975)	5,648	152	____	2.69	average age
Rubin et al (1974)	18,000	262	____	1.45	____
Lowthian (1979)	186	13	28	7.0	15.6
Woodbine (1979) Survey A	51	48	83	94	94
Woodbine (1979) Survey B	49	12	____	24	45

Figure 1.1 The incidence of pressure sores. When comparing some of the pressure sore incidence surveys one of the evident factors is that elderly people are more likely to develop a pressure sore. In general, this can be seen from the percentage of people with sores being higher when the percentage of people that are 70 years of age or older increases (Torrance 1983).

1.1.2 Location of occurrence

Pressure sores usually develop near regions of the body which have a bony prominence near the skin. In a large regional survey of the Danish population, Petersen (1975) found that less than 5% of all pressure sores were located on the upper part of the body. Figure 1.2 shows that more than 80% of all pressure sores occur at these five locations:

(1) Sacro-coccygeal region,
(2) Greater trochanter,
(3) Ischial tuberosity,

(4) Tuberosity of the calcaneus, and

(5) Lateral malleolus.

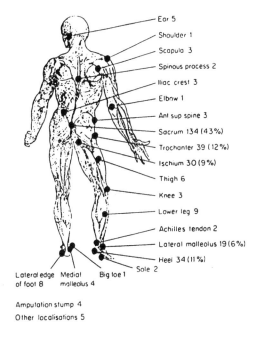

Ear 5
Shoulder 1
Scapula 3
Spinous process 2
Iliac crest 3
Elbow 1
Ant sup spine 3
Sacrum 134 (43%)
Trochanter 39 (12%)
Ischium 30 (9%)
Thigh 6
Knee 3
Lower leg 9
Achilles tendon 2
Lateral malleolus 19 (6%)
Heel 34 (11%)
Sole 2
Lateral edge of foot 8   Medial malleolus 4   Big toe 1

Amputation stump 4
Other localisations 5

Figure 1.2 A regional survey in the Danish county of Århus showed that two-thirds of all sores are found around the hips over bony prominences. Only 4% of the sores were located on the upper part of the body. The figure shows the number of pressure sores found at each location. Out of a general population of 517,000 they found 318 pressure sores (Petersen 1975).

The skin over the sacral bone is almost without muscle padding. This factor, in combination with the sacral bone having an extremely pointed shape, like that of a spur, causes pressure sores to develop there most often when an individual is positioned in the supine position (Seiler and Stähelin 1986). In the supine position, ulcers form over the sacrum and the calcaneus; in the lateral recumbent position, over the greater trochanter and lateral malleolus; and in sitting, over the ischial tuberosities.

1.2 DESCRIPTION OF TISSUES

Skin, muscle, fat, and eventually bone are all susceptible to developing a pressure sore when they are excessively mechanically deformed over a period of time. Figure 1.3 shows a cross section of each of these tissues. To understand why each tissue is susceptible to pressure, we will give a brief description of each of these tissues.

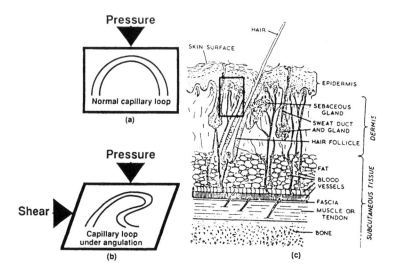

Figure 1.3 The tissue at risk from pressure. Generally, the dermis is the region that is most at risk to excessive strain from pressure. Its papillae often contain (*a*) convoluted capillary loops that are (*b*) easily bent over which results in the occlusion of blood flow. The layers are variable in size and the substructures within them are variable in number depending upon where the tissue is located in the body. (*c*) The location of the capillary loops is inside of the rectangular box located in the dermis. Adapted from (Fernandez 1987).

1.2.1 The skin

The skin is one of the larger and more versatile organs of the body. In man total skin weight is usually about 2 kg. It functions as a protective covering, retards water loss from deeper tissues, houses sensory receptors, aids in body temperature regulation, synthesizes various chemicals, and excretes small quantities of waste substances (Hole 1987). Normal skin consists of two main layers, the epidermis and the dermis.

The epidermis is the outer layer. It is avascular and consists of many layers of cells, making this tissue relatively thick in regions that are often stressed. It can be up to 1.4 mm thick in the palms of callused hands and feet; however, it is usually much thinner averaging from 70 to 120 μm for most regions. The epidermis can be differentiated into five distinct layers. The outermost layer, the stratum cornium, has about 70% less water than the innermost layer, the stratum germinativum. New cells originate in the stratum germinativum. With time they are pushed further and further outward as newer cells replace them. As they move outward they become flattened and accumulate a protein called keratin. This process, called keratinization, which causes the cells to become hardened and to die, gives the epidermis its characteristics.

The dermis is a vascular layer that lies beneath the epidermis. The surface between the dermis and the epidermis is usually uneven, because the epidermis has ridges projecting inward and the dermis has finger-like papillae passing into the spaces between the ridges (Hole 1987). These papillae increase the mechanical strength of the skin, binding the epidermis to the underlying tissues. The dermis is composed of mostly fibrous connective tissue that includes tough collagenous fibers and elastic fibers that are surrounded by a gel-like substance (Hole 1987). This also adds to the strength of skin. The dermis ranges from 0.5 mm thick in the eyelids to 3.0 mm on the soles, the average being 1.0 to 2.0 mm thick. It contains two kinds of sweat glands, the apocrine and eccrine glands. Eccrine glands are connected to pores on the surface of the epidermis. They respond throughout life by secreting sweat to cool the skin's surface when the body's temperature becomes elevated due to environmental heat or physical exercise. The apocrine glands are connected to hair follicles. They occur mostly in the armpits and groin. Other structures contained in the dermis include the lymphatics, nerves, hair follicles, smooth muscle fibers, and striated muscle fibers.

The blood vessels that supply the dermis contain thoroughfare vessels that directly connect small arterioles, called metarterioles, to the venules. The true capillaries, the site of oxygen exchange, are a shunting network of side branches of these thoroughfare vessels. The amount of blood that flows through the true capillaries is dependent upon precapillary sphincters. These are minute smooth muscle bands located on the upstream side of capillaries that constrict proportionally to the amount that they are stimulated by the sympathetic nervous system. When these sphincters are constricted, pressure inside the following capillaries decreases resulting in the collapse of these vessels. In resting tissues, most of the capillaries are collapsed, and blood flows predominantly through the thoroughfare vessels from metarterioles to the venules. Some tissues also contain arteriovenous shunts which allow arteriolar blood to completely bypass the capillaries if they are opened. These are used to help control temperature regulation. Blood vessels from beneath the skin, in the subcutaneous layer, also have branches that extend into the dermal papillae that form capillary loops. These loops are susceptible to collapse when subjected to pressure. This problem is worsened when shear is added to pressure (see figure 1.3). In the larger papillae the capillary loops can be convoluted, which further increases susceptibility to pressure and shear.

1.2.2 Fat

Directly beneath the dermis lies the subcutaneous layer which consists mostly of loose connective and adipose (fat) tissues. The collagenous and elastic fibers of this layer are continuous with the dermis except most of them run parallel to the surface of the skin. The amount of fat can vary greatly depending upon both the body region and each individual's percentage body fat. Women generally have more subcutaneous fat and it is more evenly distributed. Fat is relatively well-vascularized containing the major blood vessels that supply the skin. In general fat helps to distribute pressure applied to the skin.

1.2.3 Muscle

Skeletal muscle constitutes about 40% of the total body weight in man (Heistad and Abboud 1974). Muscle is surrounded and penetrated by layers of fibrous, avascular connective tissue called deep fascia. At rest the total blood flow through muscle is about 1 l/min; strenuous exercise increases this flow up to 25 l/min. Blood flow to muscle is determined by arterial pressure and vascular resistance in muscle. Studies in animals have shown that muscle and subcutaneous tissue are more susceptible to pressure-induced injury than the epidermis (Daniel *et al* 1981). Harman (1948) demonstrated that muscle is extremely sensitive to ischemia and that degeneration begins after 4 h of ischemia. Muscle fibers degenerate after exposure to contact pressure of 60 to 70 mm Hg for 1 to 2 h (Kosiak 1959). Muscle has a high tensile strength because of its fibers but it has a poor tolerance for compression and angulation (Torrance 1983).

1.2.4 Bone

Bones are not directly susceptible to pressure sores; however, they can also go through necrosis if they become infected by a nearby sore. An infection that can occur is osteomyelitis. Bony prominences are the usual site for pressure sores because higher pressures often result in these areas.

1.3 STAGES OF DEVELOPMENT

1.3.1 Observable patterns

When a tissue has been exposed to pressure deformation, it will react in several predictable patterns depending on how much damage has occurred.

Pressure-induced ischemia of a short duration is followed by a reactive hyperemia. Reactive hyperemia, a red flush that is observed after the release of a compressed area of tissue, is the body's attempt to correct a local metabolic debt. The area will remain red in proportion to how long the tissue was previously ischemic. This red condition is often called blanchable erythema. Compression with the finger produces total blanching of the area followed by prompt reappearance of the erythema after the finger is removed. The extra blood present will cause a small rise in temperature of the region and a slight edematous elevation of the skin. If no more insults are applied to the region, the reactive hyperemia response will lead to rapid recovery of tissue viability.

If the pressure-induced ischemia is more severe (that is, a higher pressure for the same duration or the same pressure for a longer duration) a more serious change in the vasculature called nonblanchable erythema results. This condition also follows the reactive hyperemia response. It is recognized by the erythematic skin remaining red after finger compression is removed. This is a sign that disruption in the microcirculation has occurred. In this state plasma has leaked from blood vessels into the interstitial tissues and hemorrhage has occurred (Witkowski and Parish 1982). Further deterioration of the tissue at the nonblanchable erythema site is reflected by pressure dermatitis. In this state the epidermis is disrupted and vesicles appear subepidermally. Crusts, scaling, and large vesicles may appear (Maklebust 1987). In this state, the symptoms described for nonblanchable erythema occur more frequently and are more pronounced.

The actual starting place for a pressure sore remains unclear. The earliest clinically discernible phase is blanchable erythema. Through histologic studies of 6 mm punch or scalpel samples from 59 pressure sore patients, Witkowski and Parish (1982) revealed that the earliest signs of damage occur in the upper dermis. During blanchable erythema it seems that the venules become dilated and their endothelial cells swollen and separated yielding an edemic appearance. Seiler and Stähelin (1986) observed that histologic studies consistently find perivascular round cell infiltrates, red blood cell and fibrin engorgements of the capillaries, fibrin thrombi, acute inflammatory cell infiltrates, perivascular fibrin deposits, and greatly dilated capillaries with swollen endothelial cells. The subcutaneous fat also shows signs of necrosis as early vascular changes occur. If sensory innervation is intact pain will be present. Tissue necrosis evokes an intense leukocytic response that serves to separate necrotic from healthy tissue and to produce a well-demarcated margin. The epidermis does not show any signs of necrosis until late in the process, a fact consistent with epidermal cells being able to withstand prolonged absence of oxygen both *in vivo* and *in vitro* (Crenshaw and Vistnes 1989). After necrosis of the skin, subsequent dehydration and blackening will occur. The underlying muscle is swollen and inflamed and undergoes pathological changes that are described in Husain's study (section 1.3.2). The relatively avascular deep fascia temporarily impedes downwards progress but promotes lateral extension which causes undermining of the skin. From this point, destruction will proceed rapidly with infective necrosis penetrating the deep fascia and muscle. The wound will spread along the fascial planes and bursae with joints and body cavities possibly becoming involved. At this stage, osteomyelitis can easily develop and multiple sores can communicate resulting in massive areas of tissue destruction (Torrance 1983).

In contrast to the description presented above, the results of Daniel *et al* (1981) indicate that initial pathologic changes are in the muscle and subsequently progress towards the skin if increasing amounts of pressure and/or duration of pressure are applied to an area. Daniel *et al* applied various pressures and duration of pressures to 30 Poland-China pigs and visually assessed tissue damage. They did this by using full-thickness biopsies extending from normal to affected tissue taken from the pigs 7 days after the pressure was applied. Their experimental results showed that only muscle damage occurred if high pressure for a short duration or low pressure for a long duration was applied. Muscle damage and deep dermis damage occurred if high pressure for a long duration or low pressure for a longer duration was applied. Lastly, pressures applied for a very long duration caused a full-thickness destruction of tissue including both the muscle and the skin.

Both theories could be correct when some of the variables such as the amount of pressure applied, where to the skin's surface it is applied, and the pressure thresholds before tissue damage of both the dermis and the muscle beneath it are considered. For example, pressure tends to be uniformly distributed throughout tissues between the epidermis and underlying bones. When a bony prominence is present, pressures next to the bone can be several times greater than pressures at the surface of the skin above it where the pressure is applied. If the dermis can withstand more pressure than the muscle beneath it, this suggests that a pressure sore should begin in the muscle below. These circumstances agree with the work of Daniel *et al.* However, if the dermis can withstand less pressure than the muscles beneath it, this could lead to a pressure sore beginning in the dermis. The dermis layer of the skin can become more sensitive to pressure injury due to many factors discussed later, such as the loss of collagen. This condition agrees more with the work of Witkowski and Parish. More research is needed on this topic to eliminate the numerous speculations about the progression of a developing pressure sore.

1.3.2 Tissue changes during pressure sore formation from various studies

Husain (1953) showed the effects of pressure on tissues using rats. When a pressure sore was developing, changes in the region included: loss of muscle striation, conversion of sarcoplasm into homogeneous material, and fragmentation, granularity, and eventual necrosis of the muscle fibers. The absorption of the necrosed part of the muscle fibers lead to the production of empty sarcolemmal husks. Dinsdale (1973) induced pressure sores on swine. In more severe sores he found erythrocyte thrombi and leukocyte occlusion in the venules and capillaries. Venules also demonstrated swollen mitochondria with cells having fragmented cristae. Endothelial blebs projected into the capillary lumen and swelling of the endothelial cytoplasm reduced the capillary volume to one-third of normal. Wilms-Kretschmer and Majno (1969) studied the ischemia of rats' skin and they likewise revealed the swelling and blebbing of the vasculature's endothelial cells. In addition they noted interstitial edema and endothelial damage to the lymphatics.

1.4 INSUFFICIENT TISSUE PERFUSION

1.4.1 Blood supply by the arteries

In systemic circulation blood flows from arteries, to arterioles, to capillaries, to venules, and lastly to veins. Under resting conditions the arteries contain 10%, the arterioles 1%, the capillaries 5%, and the veins 54% of the total blood volume. The other 30% is contained in the heart and pulmonary circulation.

The arteries carry high pressure blood from the heart to the arterioles. They have a peak pressure (systolic pressure) of about 120 mm Hg and minimal pressure (diastolic pressure) of 70 mm Hg during each heart cycle. Their pulse pressure, the difference between the systolic and diastolic pressures, is normally about 50 mm Hg. To do this they have strong walls consisting of three layers: an endothelial lining, a middle layer with elastic tissue and smooth muscle, and a outer layer of connective tissue. The arterioles also have three layers and have pressures of about 40 mm Hg. By the end of the arterioles, just before the capillaries, they have a mean pressure between 30–38 mm Hg with a pulse pressure of about 5 mm Hg. Their smooth muscle layer undergoes vasoconstriction or vasodilation to regulate the amount of blood flow to the capillaries. These responses are controlled by sympathetic nerves that discharge in a manner that keeps a continuous tension in the smooth muscle layer of the arterioles. The rate at which they discharge determines the amount of constriction or tone in the vessels. These nerves are controlled by the groups of neurons in the medulla of the brain that are collectively called the vasomotor center. The activity of this center is affected by many factors: blood concentrations of O_2 and CO_2, excitatory inputs from pain pathways, and inhibitory inputs from aortic baroreceptors to name a few. Ganong (1989) gives a complete description.

The capillary wall is a thin membrane made up of endothelial cells. Substances pass through the junctions between endothelial cells and some also pass through the cells by either vesicular transport, filtration, or diffusion for lipid-soluble substances. Diffusion is the most important in terms of the exchange of nutrients and waste materials between blood and tissue. Glucose and oxygen are higher in concentration in the bloodstream than in the interstitial tissues so they diffuse into them. CO_2 diffuses in the opposite direction. The rate of filtration at any point along a capillary depends upon a balance of forces called Starling forces. One of these forces, the hydrostatic pressure gradient, is the hydrostatic pressure in the capillary minus the hydrostatic pressure in the interstitial fluid. Due to these forces, substances tend to leave blood capillaries where they begin by the arterioles and enter them as they end near the venules.

The venules follow the capillaries and lead to the veins. They are similar to arterioles but have thinner walls with less smooth muscle and elastic tissue. They have pressures between 12–18 mm Hg which progressively falls to about 5.5 mm Hg in the veins. They can exhibit a vasoconstrictor response that can elevate capillary pressures. Lastly, the veins have three walls like the arteries but have much thinner walls and flaps in their lumens to prevent the backward flow of the low-pressure blood.

Figure 1.4 shows that any condition which disrupts the efficiency of the cardiovascular system described above will result in areas of tissue becoming ischemic. Torrance (1983) summarized some possible conditions such as cardiac disorders, circulatory disorders, peripheral vascular disease, anemia, and blood dyscrasias. Congestive heart failure leads to venous stasis which causes edema. Edema increases local metabolic gradients and further compromises metabolite exchange. Green (1976) mentions thrombosis of the inferior vena

Predisposing factor	Physiological response
Edema	Compromises metabolite exchange, reducing efficiency in most processes.
Congestive heart failure	Leads to venous stasis and edema.
Smoking	Has the vascular effect of constricting blood flow to the skin thus increasing ischemia.
Shock	Lowers blood pressure significantly during the first 48 h.
Anemia	The blood flow is adequate; however, the blood itself does not carry enough nutrients for tissues, especially those under pressure.

Figure 1.4 Any factor that reduces the efficiency of the cardiovascular system to deliver both oxygen and nutrients to cells of the body will also reduce resistance to pressure sore formation. Lack of oxygen and nutrients leads to ischemia and hypoxia, two factors conducive to tissue necrosis.

cava as a cause of sudden deep sacral sores. Smoking has vascular effects that cause vasoconstriction in the periphery, thus limiting the supply of nutrients to ischemic areas. Barton (1977) stated that shock can be a factor in pressure sore formation with the first 48 h being the most hazardous. Anemia and other blood disorders leave tissue devitalized thus limiting body tissues resistance to insults.

1.4.2 Waste removal by the lymphatics

One of the functions of the lymphatic system is to transport excess fluid away from the interstitial spaces and return it to the bloodstream. Figure 1.5 shows that the lymphatic pathways begin as lymphatic capillaries. These tiny tubes merge to form larger lymphatic vessels, and they lead to collecting ducts that unite with veins in the thorax. Lymphatic capillaries are microscopic, closed-ended tubes that extend into the interstitial spaces of most tissues forming complex networks that parallel the blood capillary networks. Lymphatic vessels are formed by the merging of lymphatic capillaries and have walls similar to those of veins. Their walls are composed of three layers: an endothelial lining, a middle layer of smooth muscle and elastic fibers, and an outer layer of connective tissue. Like veins, lymphatic vessels have flap-like valves preventing backward flow of lymph. Lymphatic vessels lead to specialized organs called lymph nodes that contain lymphocytes and macrophages that are vital to defend the body against invasion by disease-causing agents.

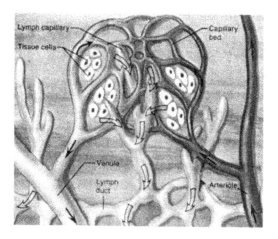

Figure 1.5 Lymph capillaries are microscopic closed-ended tubes that begin in the interstitial spaces of most tissues. Blood proteins of small molecular size and larger cellular wastes tend to accumulate in the interstitial spaces. As osmotic pressure rises in these spaces, interstitial fluid escapes into the lymph capillaries removing these accumulated substances, allowing them to be cleaned in the lymph nodes and returned to the bloodstream in the thorax (Hole 1987).

Lymph flow is due to movements of skeletal muscle, the negative intrathoracic pressure during inspiration, the suction effect of high-velocity flow of blood in the veins in which the lymphatics terminate, and rhythmic contractions of the walls of the large lymph ducts. The contractions of the smooth muscle in the walls of the lymphatic ducts are important. There is evidence that these contractions are the principal factor propelling the lymph (Ganong 1989).

Some proteins with smaller molecules leak out of the blood capillaries and enter the interstitial space. As the protein concentration of interstitial fluid rises, the osmotic pressure of the fluid rises also. This tends to interfere with the osmotic reabsorption of water by the blood capillaries. The volume of fluid in the interstitial spaces then tends to increase along with the pressure within these spaces. This increased interstitial pressure is responsible for forcing some of the interstitial fluid into the lymphatic capillaries, where it becomes lymph. The walls of the lymphatic capillaries are also permeable to macromolecules. Through this means, 25–50% of the total circulating plasma proteins are returned to the bloodstream.

The effect of pressure on the lymph drainage in tissues has been researched and considered to be a factor in pressure sore formation. Krouskop (1983) proposed that the accumulation of anaerobic metabolic waste products due to occlusion of the lymph vessels is a major factor contributing to tissue necrosis. The smooth muscle in a lymph vessel is highly sensitive to intravascular distention and to circulating humoral mediators (such as a hormone). In particular, the motility of smooth muscle is inhibited by histamine, serotonin, and prostaglandins that may be released during hypoxic states. The Wilms-Kretschmer and Majno (1969) study mentioned injury to the endothelium of lymphatic vessels that included dilation, endothelial gaps, swollen endoplasmic reticulum, and clumping of cytoplasmic contents. External pressure is seen to enhance lymph clearance until a critical closing pressure is reached. At this pressure the vessels collapse drastically reducing flow. It is not proven, but it appears more likely, that the lymphatic capillaries would close before the vessels since they are smaller and closer to the site of pressure application (Miller and Seale 1981).

1.5 CONTRIBUTION OF MECHANICAL FORCES

1.5.1 Pressure

Pressure is related to every other factor that is mentioned in this chapter. Tissues can withstand very high uniaxial pressure without causing any significant changes in cell function. However, when the uniaxial pressure applied causes ischemia by tissue distortion, a much lower tolerance to pressure is seen. The pressure duration is also important. Groth (1942), Kosiak (1959), and Dinsdale (1973) all showed an inverse relationship between tolerable pressure and time. When the magnitude of pressure and the period of time that it is applied are excessive, tissue becomes both ischemic and hypoxic. These are the two conditions under which necrosis most often occurs. Section 12.1 provides details of these studies. Many studies give the results of critical pressures to cause tissue damage measured at the skin's surface, but the pressures of interest are really within the tissue itself. Internal pressures near bony prominences are especially of interest. It would also be useful to know the specific tolerance to pressure for each type of tissue *in vivo*.

Dynamic loadings of tissue are also dangerous in tissue damage or breakdown. An example would be impact loading exerted on tissue during transfer of a patient. Hall and Brand (1979) emphasize that moderate mechanical stress may not cause pressure sores, but repetitive stress at the same level may.

1.5.2 Shear

Shear is defined as a mechanical stress that is parallel to a plane of interest (Bennett and Lee 1985). An example of shear occurs in the sacral region when a person is seated with his back at a 45° angle. The deeper fascia slides downward with the bone while the superficial sacral fascia remains attached to the dermis. This produces stretching and avulsion of the perforating arteries that supply the skin for underlying fascia and muscle (Maklebust 1987). If the skin becomes sufficiently ischemic, a pressure sore will form with wide undermining around its base (Reichel 1958). Since the microcirculation below the skin's surface is a network of small vessels oriented in all directions, it is possible that this network responds the same to shear and normal forces (pressure). Bennett *et al* (1979) showed that when enough shear is present only half as much pressure is necessary to cause vascular occlusion.

1.5.3 Friction with moisture

Four hundred grams of insensible moisture vapor flows through the skin per day (Flam 1987). Moist tissue is weaker leading to both increased maceration and excoriation. Moisture can increase the friction between two surfaces. Friction is defined as the force of two surfaces moving across one another (Krouskop 1983). Frictional forces, such as ones generated by pulling a patient across a bed sheet, can cause intraepidermal blisters and, ultimately, superficial erosions (Hunter *et al* 1974). Friction and moisture tend to produce their most damaging effects when coupled with excessive pressure. Unfortunately, this situation can occur somewhat often to elderly patients who are incontinent in extended care situations.

1.6 CONTRIBUTION OF RELATED FACTORS

Figure 1.6 shows factors which are not the direct cause of pressure sore formation but which all increase an individual's susceptibility to pressure sore development.

1.6.1 Skin temperature

Increasing the temperature by 1°C has the effect of increasing the metabolic demands of the cells and oxygen consumption by 10% in an area of tissue

(Fisher *et al* 1978). Decreasing the temperature reduces the demand of the cells but also causes a vasoconstriction that can decrease the blood supply to the area. If substantial necrosis of the tissue has occurred, the temperature of the area will be reduced due to lack of arterial blood flow to the region. Also clothing and climate are factors that can either increase or decrease the temperature of the skin. The temperature of the deeper body parts usually remains close to 37°C. The skin plays a key role in the regulation of body temperature.

Related factor	Physiological consequence
Skin temperature	Increases in temperature by 1°C increases metabolic demands by 10%.
Age	Increased stiffness and lower mechanical strength for the skin; interstitial fluid is resisted less by tissue causing cells to rupture when subjected to pressure.
Infection	Elevates body temperature, increased metabolism, and causes nutritional depletion.
Body type	People with less fat and muscle have higher pressures at their bony prominences.
Collagen formation	Factors such as stress, spinal cord injury, or old age tend to favor production of a water-soluble form of collagen at body temperature. This form is unstable decreasing the strength of the tissues.
Nutrition	Malnutrition reduces fat and muscle tissue, causing higher pressure on bony prominences
Fibrinolytic activity	When fibrinolytic activity is reduced by factors such as ischemia or friction, fibrin accumulates in the region possibly occluding blood flow.

Figure 1.6 These factors do not directly cause a pressure sore. However, when mechanical stress is applied to the tissues of a person that has one or more of these conditions, it is likely that a pressure sore will result.

1.6.2 Age

As age increases, changes take place in collagen synthesis that result in the tissues of the body having lower mechanical strength and an increased stiffness (Sacks *et al* 1985). Also, the elastin content of the soft tissue decreases, causing pressure loads to be distributed more on the interstitial fluid. The interstitial fluid acts as a cushion. If too much pressure is applied on the interstitial fluid, it will be squeezed out of a region causing cells to contact one another. This may cause cell membranes to rupture if they are stressed too greatly. After relief from the external pressure, the interstitial pressure may become sufficiently low enough to cause capillary bursting. If the resulting cellular wastes from these incidents are not removed, surrounding cells can become poisoned resulting in tissue necrosis (Krouskop 1983). Other age-related changes in the skin that increase chances of skin breakdown include the loss of dermal vessels, thinning of the epidermis, flattening of the dermal–epidermal junction, and increased skin permeability.

1.6.3 Infection

Infection will not cause a pressure sore; however, it aids in pressure sore development due to three reasons. First, infections can cause fever which increases the metabolic rate of the entire body. This increases the demand for oxygen, further endangering ischemic areas. Second, severe infection can cause nutritional disturbances and weaken the body's reserves. Third, localized bacteria increase the demand on local metabolism both by their own requirements and by the response of the body's defense mechanism (Torrance 1983).

1.6.4 Body type

A body type which distributes pressure evenly will be less inclined to pressure sore development. A thin person with little subcutaneous fat and poor muscle bulk is inclined to form a pressure sore over the bony prominences, even though low body weight is also probable. The other extreme, an obese person, has much greater weight but better padding to distribute the weight. However, these people often have poorer circulation and are more inclined to be affected by shear and friction.

1.6.5 Collagen formation

Collagen is an albuminoid that is the main supportive protein of skin, tendons, bones, cartilage, and connective tissues. It consists of two levels of organization. At the first level it has three polypeptide chains which are folded into a rod-like triple-helical molecule about 300 nm long and only 1.5 nm in diameter. The second level involves lateral and lengthwise association of the triple-helical molecules into fibrils (Prockop *et al* 1979).

When collagen is either deficient or being improperly synthesized, the tissues mentioned above become less resistant to mechanical insults. Figure 1.7 shows that collagen loss can lead to tissue destruction. Many variables which are listed as factors for pressure sore formation are often only important due to the extent that they influence collagen synthesis and degradation. One of these is emotional stress. When a person is under emotional stress, the adrenal glands increase the production of glucocorticoids and this has been demonstrated to inhibit the formation of a stable collagen trihelix. Could (1968) mentions that oxygen and trace elements are important in the synthesis of a collagen trihelix that is stable at body temperature. For example, when oxygen and trace elements such as copper are deficient the collagen formed has a very water soluble trihelix at body temperature.

It is generally accepted that the nutritional factor of ascorbic acid is necessary for normal synthesis and continued maintenance of collagen during the repair of tissue. The basic defect in collagen synthesis when it is deficient is an inhibited hydroxylation of proline and lysine to hydroxyproline amino acids that are incorporated into the collagen molecule (Robertson 1964, Stone and Meister 1962). In the absence of ascorbic acid, proline and lysine combine to form a high molecular weight polypeptide called protocollagen.

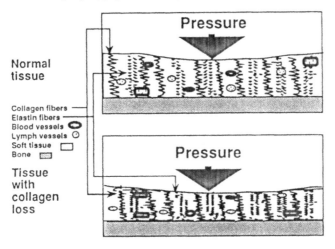

Figure 1.7 Collagen and elastic fibers act as springs to resist pressure deformation. With collagen loss the same amount of pressure will cause greater deformation and consequently more tissue destruction. Adapted from Krouskop (1983).

1.6.6 Nutrition

It has often been suggested that a person's diet can be related to resistance to pressure sores. Although it is not proven, evidence to date links malnutrition to pressure sore development. Malnutrition leads to reduction of both fat and muscle tissue, thus causing the bony prominences to have higher pressures. This indirectly predisposes a person toward the development of a pressure sore. Allman *et al* (1986) linked hypoalbuminemia due to malnutrition to the development of pressure sores in a cross-sectional survey of 634 hospitalized adult patients. Hypoproteinemia (low protein levels), through inadequate intake or excess loss causes edema, which in itself lowers resistance to pressure sore development. Ascorbic acid (vitamin C) may also have a preventive effect towards pressure sores. Taylor *et al* (1974) studied 20 surgical patients in a prospective double-blind controlled trial of ascorbic acid supplementation. He showed a significant difference in pressure' areas after only one month. The effect of nutrition on wound healing is described in Section 14.1.2.

1.6.7 Fibrinolytic activity

When a pressure sore is developing, it has been shown that a decreased amount of fibrinolytic activity is occurring. This depression in activity continues until the sore is completely healed. The fibrinolytic system contains the enzyme plasmin which is responsible for lysing both fibrin and its precursor fibrinogen. Fibrin is the insoluble protein that converts a loose aggregation of platelets which form a temporary plug into a final definitive blood clot. Fibrin deposits in and outside vessels, sometimes with intraarteriolar fibrin thrombi, may lead to complete vessel occlusion which results in tissue necrosis.

Fibrinolytic activity is affected by ischemia. Cherry *et al* (1976) observed that ischemia alone for up to 24 h had little effect on fibrinolytic activity; however, reestablishing blood flow yielded severely depressed fibrinolytic activity and increased fibrin deposits. Larson and Risberg (1977) in a 23-person study investigated how fibrinolytic activity in the skin and veins was effected by 1 to 3 h of ischemia induced by tourniquet. They had conclusions in agreement with Cherry *et al*. Immediately before blood flow was reestablished, fibrinolytic activity significantly increased. In contrast, biopsy samples taken 48 h after reperfusion of the ischemic tissue showed significantly decreased fibrinolytic activity. Turner *et al* (1969) postulated that friction at the skin's surface accelerates the depletion of the local fibrinolytic system in the dermis. Seiler and Stähelin (1986) showed that fibrinolytic activity decreased within 0 to 1.5 mm of a pressure sore border when compared with a zone 9 mm from the border. Samples taken 12 mm from the border still showed less fibrinolytic activity than normal tissue.

1.7 SPINAL CORD INJURED PERSONS

The reported annual incidence of spinal cord injury varies according to source; however, recent reports considered to be accurate indicate the annual rate is now between 30.0 and 32.1 per million persons at risk in the United States (Kennedy 1986). Mawson *et al* (1988) show that between 32 and 40% of spinal cord injured (SCI) persons develop a pressure sore within their lifetime. After severing of the spinal cord, many physiological changes take place in the body. Two well known changes are that a patient will lose both sensory perception and the ability to move in the portions of his body that have lost their nerve connections to the brain. The closer to the brain that the spinal cord is completely severed, the greater is the portion of the body rendered paralyzed. Quadriplegic persons sustain injuries to one of the eight cervical vertebrae. Paraplegic persons sustain injuries to either the thoracic, lumbar, or sacral regions of the spinal cord. Figure 1.8 shows some of the changes that cause an increased susceptibility to pressure sore development.

The first and most important change is the loss of pain sensation. When tissue, particularly the skin, is becoming ischemic, a strong sensation of pain is normally sent by C fibers along the spinothalamic tracts through the spinal cord to the thalamus of the brain. From there the impulse is sent by neurons to a somatic sensory area in the postcentral gyrus of the cerebral cortex creating the feeling of pain (Ganong 1989). This uncomfortable feeling causes an individual to move, thus relieving the pressure and restoring blood flow to the tissue. Since the SCI person does not feel pain in portions of his body that are connected to the detached portion of the spinal cord, he is unaware when tissue damage is occurring in these regions. This problem is further enhanced since SCI persons tend to move less than mobile individuals thus increasing the opportunities for ischemia due to pressure to occur.

Body characteristic	Resulting effect
Loss of sensation	Person does not perceive pain in denervated tissues, thus does not shift to remove the damaging pressure.
Reduction in movement	Since many muscles are inoperative, a person tends to move less often.
Loss of collagen	SCI persons lose collagen due to multiple reasons. This loss decreases mechanical strength of tissues.
Defective vascularity	Higher brain centers cannot regulate vascular tone. The SCI person sometimes controls this tone inappropriately causing autonomic dysfunction.
Adrenergic nerve pathways severed	The SCI person's body does not have a proper hormonal response to stress.
Unsymmetric morphology	Spinal curvature or pelvic obliquity yield unbalanced weight distributions
Abnormal soft tissue	Flaccid or spastic soft tissues yield unbalanced weight distributions

Figure 1.8 SCI persons have changes in body characteristics, which leaves them more susceptible to pressure sores.

Noble (1981) plots mean trochanteric pressure vs mean ischial pressure for 151 SCI wheelchair-bound persons. He concludes that flaccid males are expected to have a particular predisposition to pressure sores over the sitting area, followed by flaccid females and spastic males. The female with spastic paralysis is expected to present the least inherent risk.

Section 1.6 describes the importance and the formation of collagen. SCI persons form unstable collagen due to the following mechanism. The water bridges that hydroxyproline forms with other amino acids in the collagen helix are the main factor in establishing both the mechanical and thermal stability of the collagen fibrils (Traub 1974, Nemethy and Scheraga 1986). Insensate skin has a greatly decreased amino acid content per unit weight. If this interfered with the proper formation of water bridges, the skin would become weakened due to resisting mechanical insults. SCI persons have a reduced blood supply below the level of injury that could reduce the supply of nutrients to the tissues and decrease the availability of the enzyme cofactors, ascorbic acid, and molecular oxygen (Hunt 1978). The resulting localized malnutrition would alter the biosynthesis of the collagen molecule thus forming defective collagen fibrils.

Like all immobilized persons, SCI persons develop a negative nitrogen balance and catabolize large amounts of body protein. Early quadriplegic persons have increased bone turnover, mostly resorption, along with extensive collagen destruction amounting to about 270 g of collagen to each 100 mg urine hydroxyproline (Claus-Walker and Halstead 1982b). Biochemistry studies indicate that the collagen destroyed belongs to the skin as well as to the bone. Specifically, this loss is detected by a large increase in the urinary excretion of collagen metabolites such as hydroxyproline, hydroxylysine, and of glucosylgalactosyl hydroxylysine derived especially from the skin (Rodriguez and Claus-Walker 1988). This leaves the tissue more vulnerable to ulceration when subjected to excessive pressures. The hydroxylation rate of proline, expressed as the ratio of hydroxyproline to proline, is lower in the insensate skin than the normal skin for each SCI patient. This factor, in addition to the lower total content of all four amino acids, would contribute to the weakening of the skin matrix (Rodriquez and Claus-Walker 1988).

Several factors are still only speculations, but evidence points towards them increasing the risk to a SCI person of developing a pressure sore. With the spinal cord intact, many stimuli and motor reflexes are integrated by the higher brain centers to respond appropriately to total body stimuli. However, when the cord is severed both a loss of differentiation and a hypersensitivity to stimuli can develop. This is often seen as autonomic dysfunction which leads to defective vascularity (Rodriguez *et al* 1986). Also SCI persons tend to lose alpha adrenergic receptors which are responsible for mediating vascular tone and regulating blood flow in the skin. With decreased vascular tone and correspondingly decreased blood pressure, this creates sluggish venous return which increases the chance of occlusion due to pressure. Lastly, Claus-Walker and Halstead (1982a) note that the adrenal cortex and the pituitary gland of quadriplegic persons are normal and respond normally to stimulation. However, neural messages for stress are not relayed to the brain because the neurological interruptions are not sufficient to increase releasing factors and pituitary hormones, in contrast with chemical messages which are relayed by the blood. This lack of secretion of the antiinflammatory steroids may play a role in the genesis of pressure sores in the insensitive body areas.

1.8 STUDY QUESTIONS

1.1 Explain the mechanism whereby increased external pressure can increase lymph drainage.

1.2 In order of severity, list the body locations where more severe grades of pressure sores occur.

1.3 Name several physical disorders that can lead to the condition known as edema. Why does this condition leave a person more susceptible to pressure sore formation?

1.4 Describe the stages of development of a pressure sore. Which layer of skin, specifically either the dermis or the epidermis, has a greater tolerance to both pressure and shear?

1.5 Explain which vessels vary their size under neural or hormonal control and how this control changes for spinal cord injured patients.

1.6 What is the purpose of the fibrinolytic system? What response does this system assume when an area of tissue has been subjected to assault by pressure?

1.7 What are some of the likely mechanisms by which a Spinal Cord Injured (SCI) person loses collagen?

1.8 Explain at least three processes that occur at the capillary level which under normal circumstances prevent tissues from becoming ischemic?

1.9 Explain some of the anatomical microstructures of skin that provide it with strength when it is in healthy viable condition. What types of environments can weaken these structures?

1.10 Describe the role of nutritional deficiencies in pressure sore prevention.

2

Pressure Distribution in Tissue

Deng-Faa Tsay

In this chapter we introduce estimates and measurements of the pressure distribution in tissue. Several physical and computer models are included to show the results of estimating tissue pressure distribution. Experimental results suggest causes of pressure sore formation. Cushion materials affect the pressure distribution within the tissue. Reduction of blood supply plays an important role in the development of pressure sores. Model studies and experiments evaluate this reduction of blood supply. From theoretical and experimental results, we will indicate those regions at highest risk of developing pressure sores.

The following discussions are based on the assumptions that a higher pressure region has a higher probability of developing pressure sores. Some tissue may withstand high pressure for a long duration. Other factors, such as age, mobility, activity, mental state, etc., may affect pressure sores development also. However, in this chapter we will restrict discussions to pressure distributions. Later chapters will discuss other factors.

2.1 SOFT TISSUE PHYSICAL MODEL

Many researchers have worked to understand the pressure distribution inside the tissues. Model studies provide us with valuable information. However, the human body is such a complicated medium that it is difficult to find a tissue substitute that can model it satisfactorily. We must make a number of simplifications in order to develop a workable model. Several researchers have proposed different models to provide information about tissue pressure distribution. Most physical models use polyvinyl chloride (PVC) gel which has nearly the same stiffness as average tissue. PVC gel attached to a rigid core is the most commonly used model.

2.1.1 The effect of compression and shear stress

Bennett (1975) presented some simple experiments which demonstrated the effects of compressive stress, pinch shear stress, and horizontal shear stress on the water flow through tubes in his model. Figure 2.1 shows the equipment setups. He elevated a water reservoir and connected a tube through his model tissue which was a polymer with twice the stiffness as normally used PVC gel. He increased external loading and measured the flow rate of the water going through the tube.

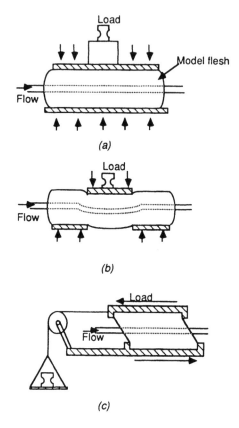

Figure 2.1 (*a*) Compressive stress is perpendicular to flow and uniform. (*b*) Pinch shear stress is perpendicular to flow at a point. (*c*) Horizontal shear stress is parallel to flow (Bennett 1975).

Compressive stress is calculated by dividing the applied load by the load area. Pinch shear stress on both sides of his model is assumed to be half of the applied load divided by the vertical cross-sectional area of the model. Bennett claims that this pressure magnitude is quite close to the actual shear stress. However, note that the gap between the bottom supports should be neither too wide nor too narrow. Otherwise, it will include other pressure effects. Horizontal shear stress is tested using two arrangements, one is with the vessel in line with the load, the other is with the vessel vertical to the load (turned 90° from that of figure 2.1(*c*)). This is called the cross shear. This pressure is the applied load divided by the load area. With these definitions of each tested pressure, figure 2.2 shows his results. The flow rate under zero loading should be equal under all tests, but figure 2.2 does not show equality. Bennett claimed that this was due to experimental error.

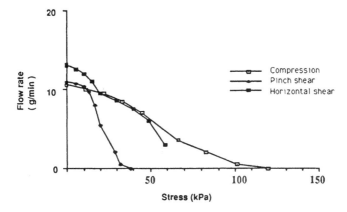

Figure 2.2 Results of Bennett's model. The water reservoir was 113 cm higher than the surface. Low pressures decreased flow rates significantly. Pinch shear stress decreased water flow the most (Bennett 1975).

The results shown in figure 2.2 suggest that the pinch shear stress is the most serious stress. Results from compressive stress and horizontal shear stress are close to each other for higher loading stresses. Compressive stress occurs when external force is applied vertically to the surface, such as the buttocks seated on a hard surface. Pinch shear stress occurs around bumpy areas, such as in the vicinity of muscle and bones. These areas have various pressure gradients around them. Horizontal shear stress can occur whenever there is a sliding action. One possible area is at the sacrum, especially when lying in bed with the back elevated. Pinch shear stress causes a drastic reduction in flow. Bennett's model is simplified yet easily understood.

2.1.2 PVC gel physical model

Figure 2.3(a) shows Reddy et al's (1982) PVC gel physical model. They marked a 2-D grid within the model to show the deformation pattern, which was recorded by camera. They compared the deformed pattern with the undeformed pattern. From the difference, they estimated the compressive and shear stress distribution within deep "tissue". Reddy et al put their model on different cushion materials and evaluated the cushions. Figures 2.3(b) and (c) show the compressive stress and shear stress inside their model when seated on soft foam. Two regions tended to concentrate shear stress and compressive stress when sitting on a cushion. They were beneath the bone core along the axis, and lateral to the core at an internal location in the buttock model. One of their purposes was to evaluate cushion design by observing deep tissue pressure distribution. They ranked cushion materials of the same thickness according to their compressive stress distribution from best to worst: (1) medium foam, (2) soft foam, (3) PVC gel, (4) viscoelastic foam, and (5) stiff foam. Then they ranked them according to shear stress distribution from best to worst: (1) medium foam, (2) soft foam, (3) stiff foam, (4) viscoelastic foam, and (5) PVC gel. They also showed that thicker cushions tend to reduce stress and distribute it more uniformly.

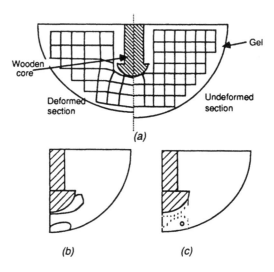

Figure 2.3 Reddy *et al*'s PVC gel buttock model with 2-D grid to show the deformation. (*a*) Cross-sectional view of the model. The right half is undeformed, the left half is deformed to show the contrast. (*b*) Compressive stress distribution when sitting on "soft" foam. The contour shows the 4-kPa isostress lines. (*c*) Shear stress distribution on the same material. Broken lines are 2-kPa isostress lines and solid lines are 2.7 kPa (Reddy *et al* 1982).

These two models show us that not only is compressive stress important, the shear stress is at least as important or even more important for pressure sore development.

2.1.3 Model based on blood flow change

Other researchers believed observing changes in skin blood flow might provide more valuable information. Some of them believed the skin blood flow change is less influenced by pressure loading than tissue distortion.

Sacks *et al* (1985) developed their model of skin blood flow changes and tissue deflection by using cylindrical indentors. They assumed the load and bone were cylindrical. They claimed that the skin blood flow changes were influenced by three factors: the ratios of bone depth, the ratios of indentor diameter to bone diameter, and percentage compression of the tissue overlying the bone. They proved that the load was unimportant. Measurements of these three factors are more important than measuring pressure distribution. They also pointed out that measurements based on percentage change of blood flow as a function of the displacement of bone depth and bone diameter were better than those based on blood flow as a function of applied pressure.

Figure 2.4 shows the dynamic mechanical model of Kett and Levine (1987). It provides information about tissue deflection due to applied force and skin recovery after pressure relief. They proved that tissue blood flow is better related to tissue distortion than externally applied pressure. All parameters of this model are derived based on a cylinder of soft tissue 35 mm in diameter by 40 mm long. This model included bone soft tissue, blood vessels, lymph vessels, and seat cushion. Elastic elements $K1$, $K2$ and resistive element $B2$ represent the viscoelastic property of the model. Resistive element $B1$ results from the bodily fluid forced in and out of the model by external pressure. They did not include cushion properties in their first studies. Elastic elements $K1$ and $K2$ are derived from the indentor model of Sacks *et al* (1985). These two elements are nonlinear. Resistive element $B1$ represents the interstitial fluid flow out of the tissue. Flow $V1$ is the fluid source. The value of this parameter is derived from the relation among fluid exchange to tissue permeability, capillary diameter, capillary length, and blood viscosity. The displacement function then

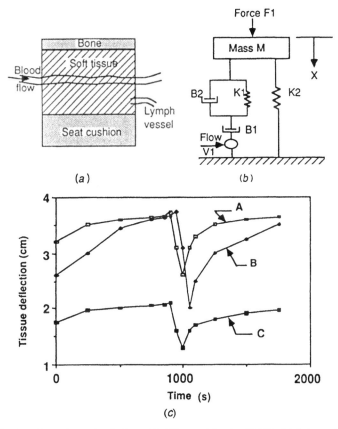

Figure 2.4 The dynamic mechanical model of Kett and Levine (1987). (*a*) Schematic model with cushion included. (*b*) Dynamic model. where *M* is the mass. *F*1 is the external test force. (c) The effect of performing pressure relief. Curve A: 0.5-N force, 30-s relief. Curve B: 0.5-N force, 60-s relief. Curve C: 0.4-N force, 30-s relief.

directly influences the blood supply. Figure 2.4(*c*) shows some interesting results from this model. By performing longer pressure relief, the tissue then takes longer to come back to its compressed state. Therefore, the blood supply can survive longer. Also, smaller external pressure (Curve C) yields smaller displacement. Thus the conclusion is applying longer pressure relief or lower pressure will reduce the possibility of pressure sore development.

These two models show us that applied pressure is not the critical parameter for pressure sore formation. The tissue distortion is more important. Where tissue distortion occurs suggests the location where pressure gradient occurs. That is where the shear stress occurs. Thus, though all these models approach the problem in different ways, they are actually related.

2.2 COMPUTER MODEL

Computers simplify complicated analyses. The finite element method (FEM) has been applied to many types of structures. This nondestructive method can be used to estimate stresses and strains inside the structure. While the simulation processes are useful in predicting the stress–strain states and addressing the high risk of internal tissues, we should use caution as most computer models are based on an elastic model with small displacement. Improved modeling will use advanced finite element modeling (nonlinear element, large displacement) and noninvasive imaging (CT, MRI, ultrasound) as in Section 10.1.3.

2.2.1 Stress, strain, and shear

One dimensional element
Before going into the model itself, we will explain stress, strain, and shear. Starting from a one dimensional element, figure 2.5(*a*) shows the deformation of this axial element. Stress σ_x is defined as the load divided by the cross-sectional area of the element. Strain ε_{ex} is defined as the displacement per unit length. The stress–strain relation is:

$$\sigma_x = E\varepsilon_{ex}$$

$$\varepsilon_x = \varepsilon_{ox} + \varepsilon_{ex}.$$

The strain–displacement relation is:

$$\varepsilon_x = \frac{du}{dx}$$

where σ_x is the elastic stress (Pa), E is Young's modulus (Pa), ε_{ex} is the elastic strain (m/m), ε_{ox} is the initial strain due to thermal effects. For a built-in strain, ε_x is the total axial strain, and u is displacement (m). Young's modulus is a property of the elastic element. It defines how easily a material can be lengthened or squeezed in the longitudinal direction. If we twist this element, we will have shear stress. Shear stress is more apparent in 2-D and 3-D elements.

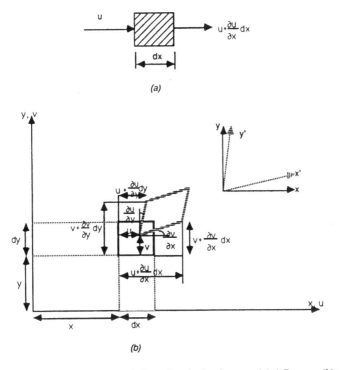

(a)

(b)

Figure 2.5 Strain–displacement relation of a single element. (a) 1-D case. (b) 2-D case (Robinson 1981).

Two dimensional element
Figure 2.5(b) shows the deformation of a membrane element. The strain field includes three elements:

$$\{ \varepsilon \} = \{ \varepsilon_x \; \varepsilon_y \; \gamma_{xy} \}$$

where ε_x is the total direct strain in the x direction, ε_y is the total direct strain in the y direction, and γ_{xy} is the total shear strain. The strain–displacement relation is given as:

$$\varepsilon_x = \frac{\partial u}{\partial x}$$

$$\varepsilon_y = \frac{\partial v}{\partial y}$$

$$\gamma_{xy} = \frac{\partial v}{\partial x} + \frac{\partial u}{\partial y}.$$

For this two dimensional element, the stress–strain relation must include Poisson's ratio v to describe the vertical deformation due to horizontal stretch. This relation is given by the matrix form:

$$\begin{bmatrix} \sigma_x \\ \sigma_y \\ \gamma_{xy} \end{bmatrix} = \frac{E}{(1+v)(1-2v)} \begin{bmatrix} (1-v) & v & 0 \\ v & (1-v) & 0 \\ 0 & 0 & (1-2v)/2 \end{bmatrix} \begin{bmatrix} \varepsilon_{ex} \\ \varepsilon_{ey} \\ \gamma_{exy} \end{bmatrix}.$$

As in the 2-D case, initial strains are included in Poisson's ratio, where v has units of m/m.

Three dimensional element
With same basic idea and more complicated formulation, the 3-D case will be discussed below.

2.2.2 Finite element method model

A FEM model replaces the actual structure in a mathematical representation. This replacement makes software analysis possible for a complicated structure. Basically, a material is partitioned into finite elements. Each element is assumed to have isotropic parameters. A linear or piece-wise linear relation is applied within elements and between elements.

Problems in using the FEM in the human body arise due to its complexity. Chow and Odell (1978) made a 3-D FEM model to study deep tissue pressure distribution and the effect of seating environment. Their model was a 100-mm radius hemisphere with a rigid core at center. The "soft tissue" was assumed to have a constant Poisson's ratio of 0.49, and a constant Young's modulus of 15 kPa. Figure 2.6(a) shows how the model was axisymmetrically partitioned into 33 elements. They examined two kinds of pressure. One is hydrostatic pressure σ_h which they believed did not deform the tissue or decrease the volume, and thus was less important. The other is octahedral shear stress σ (or von Mises stress) which they believed contributed to tissue deformation and was more important. Hydrostatic pressure is an invariant quantity in the strain tensor for a given state of stress. Von Mises shear stress is the maximal shear stress before the material yields. The elastic failure takes place when the shear strain energy per unit volume at a point is equal to the shear strain energy per unit volume in a specimen of the same material in the simple uniaxial test (Ross 1987).

Chow and Odell applied the following procedures to derive the pressure distribution inside their model:

(1) The surface pressure distribution P is divided into incremental surface pressures ΔP_i where

$$P = \sum_{i=1}^{n} \Delta P_i \tag{2.1}$$

and n is the number of load increments.

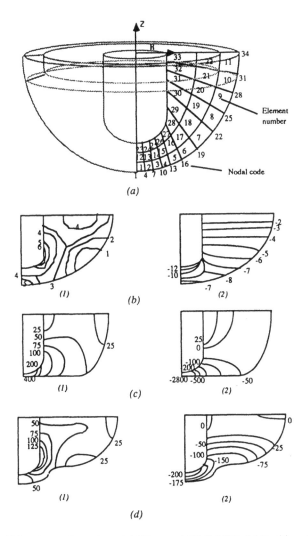

Figure 2.6 Finite element buttock model (Chow and Odell 1978). (*a*) Partition of the model. Each element is axisymmetric. (*b*) Model just beneath water. (1) von Mises stress, (2) hydrostatic stress. (*c*) Model on a flat frictionless rigid surface. (1), (2) same as (*b*). (*d*) Model subjected to a modified cosine pressure distribution (1), (2) same as (*b*).

(2) Convert incremental surface pressure into surface nodal forces, ΔF_j, by boundary integration. The total vertical forces should be equal to the applied force.

(3) The stiffness of each element is calculated by its coordinates, Young's modulus, and Poisson's ratio by integrating from the incremental displacement–strain matrix $[B]$, and stress–strain matrix $[D]$ (Robinson 1981) as follows:

$$[K]_{\text{element}} = \int_{\text{vol}} \int \int [B]^T [D] [B] \, d(vol). \qquad (2.2)$$

The stiffness matrix describes the relation between the applied force and the displacement. In the 1-D case, the spring constant K is the stiffness. It relates the force and displacement in a simple formula $F = Ku$. In the multidimensional case, it is in matrix form $\{F\} = \{K\} \{u\}$ (Burnett 1987).

(4) The element stiffness matrices are then transformed into a global stiffness matrix $[S]$

(5) The incremental displacements $\{\Delta u\}_i$ are calculated from the known incremental nodal forces $\{\Delta F\}_i$ as follows:

$$\{\Delta u\}_i = [S]^{-1} \{\Delta F\}_i. \qquad (2.3)$$

(6) The nodal incremental displacements $\{\Delta u\}_i$ are then converted to incremental strains $\{\Delta \varepsilon\}_i$, and then strains are converted into incremental stresses $\{\Delta \sigma\}_i$ for each element

$$\{\Delta \varepsilon\}_i = [B] \{\Delta u\}_i \qquad (2.4)$$

$$\{\Delta \sigma\}_i = [D] \{\Delta \varepsilon\}_i. \qquad (2.5)$$

(7) The stress vector $(\Delta \sigma_r, \Delta \sigma_z, \Delta \sigma_\theta, \Delta \tau_{rz})$ due to the incremental load is then added to the cumulative stress vector $\{\sigma_{\text{cum}}\}$ at each node and each quadrature integration point.

$$\{\sigma_{\text{cum}}\}_i = \{\sigma_{\text{cum}}\}_{i-1} + \{\Delta \sigma\}_i \qquad (2.6)$$

where $\{\sigma_{\text{cum}}\}_{i-1}$ is first corrected for rotation and change in area.

(8) All stresses derived so far are based on the same coordinate system. We can transform these stresses along their principal axis such that there will be no shear stresses along that direction. These are principal axes of stress. The normal stresses along these axes are principal stresses (Burnett 1987). Figure 2.5 shows the existence of this axis. Shear stress usually is maximal at 45° to the principal axis. The von Mises stress and hydrostatic stress are calculated as follows:

$$\sigma = \rho\{([(\sigma_1 - \sigma_2)^2 + (\sigma_2 - \sigma_3)^2 + (\sigma_3 - \sigma_1)^2]/2)\}^{1/2} \qquad (2.7)$$

$$\sigma_h = (\sigma_1 + \sigma_2 + \sigma_3)/3 \qquad (2.8)$$

σ_1, σ_2, and σ_3 are three principal stresses transformed from the stress vectors $(\Delta \sigma_r, \Delta \sigma_z, \Delta \sigma_\theta, \Delta \tau_{rz})$.

(9) The geometry of the model is then adjusted by adding the displacements $\{\Delta u\}_i$ to the coordinates $\{r\}_i$ and $\{z\}_i$

$${r}_{i+1} = {r}_i + {\Delta u_r}_i \tag{2.9}$$

$${z}_{i+1} = {z}_i + {\Delta u_z}_i . \tag{2.10}$$

(10)Using these new coordinates (${r}_{i+1}$, ${z}_{i+1}$), the procedure is then repeated.

The surface pressure distribution is derived for different loading situations. When the model is submerged just beneath the surface of the water, the pressure distribution can be calculated as

$$P = \rho gh = -9.807z \quad \text{(Pa)}$$

where ρ is the density of water, g is the gravitational constant, and h is the depth beneath the surface of the water. Figure 2.6(b) shows the von Mises stress distribution and hydrostatic stress distribution in this case. Since the z axis was chosen positive upward, we see negative hydrostatic stresses. However equation (2.7) shows that von Mises stress is always positive. Tissue deformation is almost negligible here. If the model were put 907 mm beneath the surface of the water, the model surface pressure distribution would be

$$P = 8897 - 9.807z \quad \text{(Pa)}.$$

The von Mises stress is almost the same as given above. Hydrostatic stress is higher but with the same increments. Deformation still can be ignored. This model actually proved the benefits of using a water bed for pressure sore prevention. Even when the model was floated in mercury, it still had uniformly distributed von Mises stress and very small deformation. Since patients will sink deep into a water bed, some researchers have suggested using a high density liquid in place of water. There are good reasons for this substitution.

When the model was seated on a flat frictionless rigid surface, it showed large deformation at the contact area (figure 2.6(c)). Conry and Seireg (1971) proposed a method of deriving deformation at the contact area. This method was applied to Chow and Odell's model. In order to find the pressure distribution when the model was seated on a cushion, a modified cosine pressure distribution was used so that at the contacts all external forces are vertical to the surface. We have seen that the von Mises stress was not uniformly distributed. If seated on a 4-inch (100-mm) foam cushion, the von Mises stress distribution is similar to that in figure 2.6(d), which shows that shear stresses are much larger inside the model than at the surface. When seated on a flat rigid surface, the region along the vertical axis experiences the highest compressive and shear stresses. When seated on a "cushion", this region has the highest compressive stresses. However, the region lateral to the bottom of the bone experiences the highest shear stresses. Shear stresses along the vertical axis are quite small and uniform. The reason is that the bottom of the model is merged into the cushion, therefore, even if it has very high compressive stresses, it has low shear stresses. At the region just above the cushion, it has a large gradient in pressure, therefore, even though the compressive stresses are lower, the shear stress are higher.

2.3 DEEP PRESSURE FROM SKIN PRESSURE

In the last two sections, all models show us the importance of determining deep tissue pressure distribution. However, there are difficulties in measuring deep tissue pressure directly. Noninvasive measuring techniques, such as using ultrasound and magnetic resonance imaging, show anatomy, but do not show the correct deep tissue pressure. Invasive methods may change the pressure distribution when taking measurements and human subjects are not easily available for the tests. However, efforts have been made to explore deep tissue pressure distribution *in vitro* and *in vivo*. They have provided direct information and have supported the reliability of the models.

2.3.1 Deep tissue pressure in animals

The first in-depth measurements were made by Le *et al* (1984) *in vitro* and *in vivo* on their model and animal subjects. They used monolithic silicon pressure sensors with a sensing needle inserted into tissues. Chapter 9 gives a more detailed description of this method. Measurements were made at first *in vitro* in a "rump roast" using plumbing pipe as a bone model. *In vitro* experiments proved this was a feasible method. *In vivo* experiments were made on a Yorkshire pig. The needle was inserted vertically into the selected area.

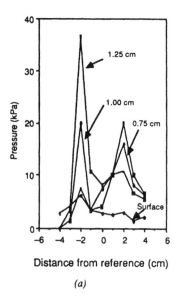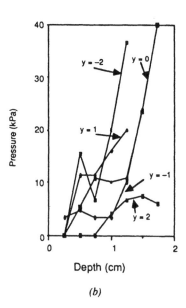

Figure 2.7 (*a*) Pressure distribution sampled at different horizontal planes in the tissue underlying the left trochanter and ischium and pressure measured at the surface. (*b*) Pressure as a function of the sampling point at various coordinates under the left trochanter (Le *et al* 1984).

Measurements were taken by changing the needle depth. An air bladder at the surface monitored the skin pressure. Figure 2.7(a) shows the deep tissue pressure may be as high as 275 mm Hg while the surface pressure is less than 50 mm Hg. The pressure near the bony prominence is nearly three to five times higher than the skin pressure. Figure 2.7(b) shows pressure changes with respect to the depth for different coordinates. It shows that the pressure increases from skin to dermis, then decreases, and then increases again in the muscle layers. The decrease was believed to be measurement artifact.

Le et al's experiments illustrated that the skin pressure distribution does not reflect the deep tissue pressure distribution. However, they did not show the tissue pressure distribution due to different loading situations, such as sitting on a well-designed cushion. Also, they did not separate shear stress and compressive stress. They measured the scalar pressure. It would be difficult to try to do measurements for different loading situations in vivo.

Did Le et al measure the compressive stress only? We cannot separate shear from compressive stress using these scalar values. They should include both kinds of stresses. Chow and Odell's FEM model in figure 2.6(d) yields a von Mises stress of 125 mm Hg near the bone and 50 mm Hg at the contact surface. The magnitude of hydrostatic stress is dependent on the loading area. The results of this model fit well the experimental measurements. However, the model can only be derived under certain restrictions. Therefore we must use caution when applying these models.

2.4 DEPTH WHERE PRESSURE SORE ORIGINATES

All the models try to prove that either higher pressure or tissue deformation causes pressure sores. These studies are based on the assumption that the tissue is an isotropic material. However, we will learn from other chapters that this assumption may be too simplified. But these models and experiments are valuable for their explanation of the inner pressure distribution and how this affects the blood flow. Relative pressure causes relative danger for pressure sores. From an engineering viewpoint, we believe pressure sores start in high-pressure regions.

2.4.1 Pyramid structure: compressive stress

The supporting interface, soft tissue, and bony prominence form a pyramidal structure. All the interface pressures tend to concentrate at the top of the pyramid, next to the bone. Even without the models, our experience tells us there are higher pressures near bony prominences. Although tissue complexities make pressure sore formation variable, high-pressure regions are more at risk for ulceration.

High pressure between hard tissue and the contact interface certainly blocks the blood supply and lymph drainage. Chapter 1 describes the reasons for pressure sore formation. Compressive stress is the easiest way of observing these high-pressure regions. Model studies show that compressive stress is almost vertically directed from contact interface to the bony prominence.

However there are other stresses besides compressive stress. Simply redistributing pressure under contact interfaces does not eliminate risk of pressure sores.

2.4.2 Beyond pyramid structure: shear stress

Shear stresses are more important than compressive stresses. Unfortunately, they are not as observable as compressive stresses. Moreover, they are much higher within the deep tissue than at the surface. The highest shear stress is at the largest pressure gradient within the tissue. The interface between hard tissue and soft tissue is the most susceptible area, such as in the vicinity of bone and the junction between two bundles of muscle. Also, shear stress reduces blood flow more easily than compressive stress. Therefore, pressure sores likely begin in the deep tissue. When they are observed at the skin, they may have developed first near the bone and later in the skin.

Later chapters will describe methods of measuring pressure at the surface. Measuring deep pressure distribution directly is difficult. However, Chow and Odell's model gives us confidence that if we know the surface pressure distribution and all the tissue parameters, we can estimate deep tissue pressure distribution. With the aid of modern computers, it should be possible to combine the surface pressure measurements and the models to yield estimates of deep pressure distribution.

2.5 MEASURING SKIN BLOOD FLOW

Some researchers believe that blocking the blood flow is the major cause of pressure sores. Houle (1969) proposed that the maximal seating interface pressure should not exceed 32 mm Hg which is the capillary blood pressure. However Landis (1930) shows that capillary blood pressure has a large variability in the finger and varies with posture and hyperemia. It likely also varies with site. Holloway *et al* (1976) showed that skin blood flow reduces with pressures above 30 mm Hg, but does not stop until a much higher pressure is reached. Sacks (1989) made a theoretical prediction of a pressure–duration curve for avoiding pressure sores. Bennett (1975) based his models on the belief that reducing the blood supply caused pressure sores. The model of Kett and Levine (1987) is also based on evaluating blood flow. High pressures, compressive or shear stresses, and tissue deformation certainly all affect the blood supply.

2.5.1 Available methods

There are several ways to measure blood flow. Indicator dilution is the most commonly used invasive method. We inject an indicator into a blood vessel through a catheter and record the indicator concentration to measure the blood flow. Ultrasonic flowmeters are the most commonly used noninvasive instruments (Webster 1978). Watkins and Holloway (1978) measured blood flow by observing the Doppler shift of laser light. All these techniques are sophisticated and widely used.

Doppler shift

However, these methods are not successful for measuring blood flow under external loading. Sableman *et al* (1989) applied laser Doppler shift techniques to measure skin blood flow and found the limitations of this method. Large loading by a 32 mm diameter hemispherical indentor reduces the blood flow dramatically. Errors become significant as loading increases and blood flow reduces. They made some improvements in order to measure skin capillary blood flow under very light pressure. Data are measurable when load is higher than 30 g (10 mm Hg). Blood flow decreases as load increases up to 100 g (33 mm Hg), but not significantly.

$^{81}Rb/^{81m}Kr$ ratio

Gross *et al* (1982) measured real time blood flow under external loading by determining the $^{81}Rb/^{81m}Kr$ ratio. Rb-81 decays to Kr-81m in the cells. Both of the elements give off high-energy gamma rays. The instantaneous ratio of Rb-81 to Kr-81m activity should reveal the blood flow change under loading. However, in their *in vivo* experiment, no significant change could be observed.

Noninvasive measurement methods can be applied only to measuring surface blood flow. Deeper blood flow can be measured only invasively. Meanwhile, all noninvasive measurements produce errors at low blood flow rates. Since pressure sores likely start from deep tissue, simply measuring skin blood flow changes does not provide valuable information. However, we can apply this method to measuring blood flow changes in fingers and toes, since tissue at these areas is thinner.

2.6 PRESSURE DISTRIBUTION AT DIFFERENT LOCATIONS

Skin surface pressure distribution is strongly dependent on body posture. When sitting vertically, the ischium experiences the highest pressure. When lying supine, the sacrum experiences the highest pressure. When lying on the side, the trochantor experiences the highest pressure. Lindan (1961) used a "nail bed" to measure the surface pressure distribution. Fernandez (1987) also indicated the locations most susceptible to pressure sores. We have noted that skin pressure distribution does not reflect deep tissue pressure distribution. However, locations with the highest skin pressures are also expected to have the highest internal pressures.

Model studies have shown that shear stresses are higher with increased pressure gradient. Figure 2.8 shows three areas with possible highest shear stress. From Lindan's studies, for a normal person, sacral pressure is as high as 50 mm Hg when lying supine, and ischial pressure is 100 mm Hg when seated with the feet supported. Deep tissue pressures would be higher.

Fat persons will have more uniform pressure distributions than others, but they may experience another risk. An older person with loose skin may have this risk also. Figure 2.9 shows that folded skin is created by surface shear at the interface. The folded skin experiences high compressive and shear stress and poor circulation. These areas have high risk of pressure sores, which start in the skin. Therefore, they are easily detected and prevented.

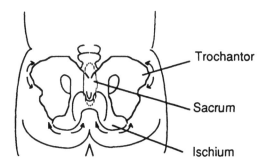

Figure 2.8 Part of human skeleton and areas experiencing high shear stresses. Stress distribution depends on body postures. Arrows indicate possible shear stress directions.

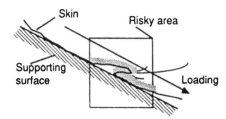

Figure 2.9 When a person is fat or has loose skin sliding along the supporting surface, they may have their skin folded. The risky area indicates the most susceptible area for pressure sores.

2.7 STUDY QUESTIONS

2.1 Sketch a situation in which a patient might encounter pinch shear.

2.2 Describe how Reddy *et al* measured pressure distribution and shear inside the deep tissue of their PVC gel buttock physical model.

2.3 Sketch the displacement vs. time for a step increase in force for the dynamic mechanical model of Kett and Levine. From identified mechanical elements, give the equation for the time constant τ.

2.4 Define and give units for Young's modulus and Poisson's ratio.

2.5 Describe how to relate mechanical parameters to the equation used an a single element of a finite element model.

2.6 Calculate the surface pressure distribution when the model of Chow and Odell is floating on mercury (density of mercury = 13.6 g/cm^3).

2.7 Give reasons why we might select liquid-filled supporting materials.

2.8 Sketch the placement of hypodermic needles and silicon pressure sensors for measuring deep tissue pressure.

2.9 Explain why you believe most pressure sores originate in the skin or in deeper tissue.

2.10 Describe the advantages and disadvantages of different methods of measuring skin blood flow.

2.11 State the location on the body where the highest pressure has been recorded. Why is it highest there?

2.12 Describe the laser Doppler method of measuring skin blood flow.

2.13 Describe the Rb-81, Kr-81m method of measuring skin blood flow.

3

Signs of Pressure Sores

Basel Taha

This chapter presents the information necessary to detect early warning signs of pressure sores and to reliably identify persons most at risk of developing them. We start with a description of skin appearance during and after pressure application. We then present risk factors that contribute to the process of pressure sore development and their usage in assessment tools designed to predict the risk of contracting pressure sores. In the last section we discuss some objective measurement techniques used to evaluate various physical and physiological changes associated with pressure sore development.

3.1 EARLY WARNING SIGNS

The human skin has a variety of colors and shows remarkable individual variations even within racial groups. The appearance of the skin is partly due to the reddish pigment in the blood of the superficial vessels. Mainly, however, it is determined by melanin, a pigment manufactured by dendritic cells called melanocytes, found among the basal cells of the epidermis. Their numbers in any one region of the body are roughly the same within and between races. Color differences are due solely to the amount of melanin produced and the nature of the pigment granules. There are two kinds of melanin: dark brown eumelanin and pale red or yellowish phaeomelanin. The redness of white-skin people is therefore dependent on the amount of blood in the skin capillaries and its oxygenation (Goller *et al* 1971).

Skin temperature, on the other hand, is governed by the blood flow through it as opposed to its amount (Goller *et al* 1971, Barnes 1967). At any moment, the skin's temperature is a function of the amount of heat flow to it from within the body and the amount of losses or gains through the surface by conduction, convection, and infrared radiation.

Since changes in skin color and temperature due to pressure aid in predicting and consequently preventing pressure sores, we will describe below early visual and thermal responses of the skin to pressure application.

3.1.1 Ischemic pallor

When enough pressure is applied to the skin of a white subject through a transparent surface, a sharply outlined spot that looks much whiter than the rest of the skin is immediately noticed at the pressure site (figure 3.1). This whiteness is due to the reduction of blood content in the skin and is therefore called an ischemic pallor. Sachs and Miller (1967) define blanching or ischemic pallor as "the maximum or near maximum decrease in normal redness of the skin as can be produced by applying sufficient pressure." They determined the pressure necessary to produce an ischemic pallor under the sacrum to be between 60 and 88 mm Hg.

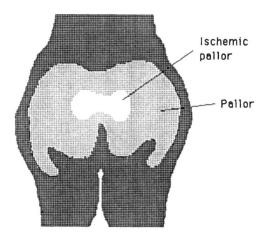

Figure 3.1 A person lying in the supine position on a transparent hard surface develops an ischemic pallor under the sacrum. Notice that skin is whiter than normal over the entire contact area but maximal whiteness is apparent in the area of highest pressure.

An observable whiteness (pallor) can be produced under light pressure on some parts of the body like the fingers and it disappears quickly when the pressure is relieved (Miller and Sachs 1974). This suggests that the capillary loops in the dermal papillae—being the most superficial vessels in the skin—collapse, effectively squeezing blood out of them and blocking further blood flow (figure 1.3). Higher pressures that occlude deeper arterioles and venules would increase the blanching and lead to the formation of the ischemic pallor.

Ischemic pallor is the first visual sign of change in the skin characteristics due to pressure. However, it is hard to notice since it disappears quickly after the pressure is removed as blood rapidly flows back to the deprived area of the skin. Goller *et al* (1971) argue that, in general, the appearance of a pallor on the skin does not necessarily imply total deprivation of blood in the skin. They state that pallor of the skin indicates only that the capillaries are empty, a situation that may arise when blood is passing rapidly from the arterioles to small veins through arterial shunts (Section 1.2.1).

Pye and Bowker (1976) set up a simple model that shows the reduction in blood flow and consequently in heat flow due to pressure. In order to verify the model they applied pressure via a perspex (Plexiglas) indenter to various areas of soft tissue while recording the changes in skin temperature using thermography (Section 3.4.1). They reported an initial decrease in temperature right after the application of pressure but this effect quickly disappeared to leave a normal thermograph again.

3.1.2 Reactive hyperemia

Blood supplying nutrients to the skin tissue is subject to the same local autoregulatory mechanism for blood flow as that found elsewhere in the body. Therefore, when the skin tissue is deprived of nutrients because of a completely blocked blood flow for some time, the skin blood flow becomes far greater than normal immediately after release of the pressure, causing the skin to have a red blush called reactive hyperemia (Guyton 1981). During reactive hyperemia, the visible red flare becomes very intense in the first few minutes and then gradually diminishes (Trandel 1975). Reactive hyperemia is a normal reaction that supplies oxygen and nutrients and washes away accumulated waste (Miller and Sachs 1974) and should normally disappear within 10 to 15 min after pressure removal, otherwise underlying tissue damage is assumed (Gosnell 1987). Light finger touch will cause blanching of this erythema, indicating that the microcirculation is intact (Waterlow 1985).

Levine et al (1989a) state that in addition to distinct erythema, reactive hyperemia is usually accompanied by warmth of the skin due to increased blood flow in the dermis. However, Goller et al (1971) suggest that there may be no correlation between the color and the temperature of the skin. The color of the skin is dependent on the amount of blood in the capillaries and its oxygenation while skin temperature depends on blood flow through it. Mahanty and Roemer (1980) investigated the dependance of the peak temperature rise and the time of its occurrence on the magnitude and duration of pressure application. They found that peak temperature rise increased as the pressure magnitude or duration increased. They repeated the study (Mahanty et al 1981) on paraplegic subjects to compare their thermal responses to those of able-bodied persons (figure 3.2). They found close similarity in results for paraplegic and able-

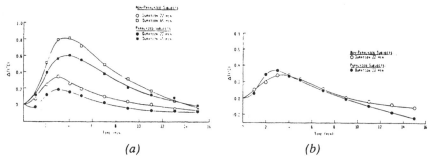

(a)   *(b)*

Figure 3.2 The peak temperature rise associated with reactive hyperemia increases with increase in pressure magnitude or duration (Mahanty 1981). (a) 200 mm Hg pressure on trochanter. (b) 300 mm Hg pressure on trochanter.

bodied subjects and explained it by citing studies that emphasize the independence of reactive hyperemia of innervation. Section 3.3.1 presents some findings, using thermography, of thermal responses associated with reactive hyperemia.

Sachs and Miller (1967) found that reactive hyperemia corresponds with the outline of ischemic pallor and that these outlines also correspond to the outline of the developing pressure sore. Thus, it is clear that the reactive hyperemia reaction is an important early warning sign of skin breakdown. It is usually relatively easy to detect visually. This is not the case however in the elderly where degeneration of small vessels in the dermis may mask the reactive hyperemia stage due to reduced local blood flow (Levine et al 1989a).

3.1.3 Stages of pressure sore development

Figure 3.3 lists the stages of pressure sore development along with a description of each stage's appearance of the skin and amount of tissue involvement. Some pressure sores, called closed pressure sores, involve deep damage to tissue while only showing a small ulcer on the surface (Waterlow 1985).

Stage	Description
1 epidermis	A circumscribed reddened area; may appear bruised, but there is no break in skin integrity; light finger pressure causes blanching.
2 dermis	A break in the skin surface, with exposure of the dermis and reddened area surrounding; serous drainage may be present; if sensory innervation is intact pain is present.
3 subcutaneous tissue	An ulceration involving subcutaneous tissue; serous or purulent drainage may be present; ulcer edges are distinct and surrounded by erythema and induration.
4 subcutaneous fat	Lesion extends into subcutaneous fat; small vessel thromboses and infection compound fat necrosis; deep fascia temporarily impedes downward progress.
5 muscle and bone	Necrosis penetrates deep fascia and muscle induration progresses rapidly; joints and body cavities can become involved; multiple sores may communicate.

Figure 3.3 Pressure sore development stages are classified according to degree of tissue involvement. Note that a stage 1 pressure sore includes the normal reactive hyperemia. Adapted from Gosnell (1987) and Waterlow (1985).

3.2 PRESSURE SORE RISK ASSESSMENT

Success in the prevention of pressure sores is directly dependent on the skills in pressure sore assessment. Gosnell (1987) divides the pressure sore assessment procedure into two parts: integumentary assessment and risk factor assessment. Several pressure sore assessment instruments have been designed using factors from both parts to give a quantitative measure of the risk of contracting a pressure sore. In this section we present observations and factors that indicate potential for forming pressure sores.

3.2.1 Integumentary assessment

Assessment of the integumentary system should be done with consideration for each of the vital functions of the skin outlined in Section 1.2.1. Figure 3.4 is a guide to skin assessment factors. A large amount of integumentary assessment information can be gathered by inspection of lesions, pressure areas, blemishes, and color. Special attention should be given to areas of the skin where skin surfaces approximate and over bony prominences. Figure 1.2 shows the common pressure points.

As mentioned earlier, the first important early warning sign of skin tissue insult is reactive hyperemia. A normal reactive hyperemia should disappear

Function	Focus	Findings
Temperature regulation	Palpate skin surface, including extremities; observe color	A warm flesh pink tone is normal. Hot suggests increased blood flow as a result of the body's response to inflammation or infection, and the surface usually is red or purplish in color. Cool or cold is usually due to decreased blood flow to the surface as a result of impaired circulation due to either internal or external factors; color is usually white or pallor.
Sensory communicator	Temperature, tactile, and two-point discrimination on various parts of the body should be checked	Normally one should be able to distinguish degrees of temperature, sharp, dull and pressure sensations against skin surface. Diminished sensation may be generalized or affect only a part of the body, such as lower extremities.
Storage of water and fat	Observe skin tone (turgor and tension) and body build	Skin should normally be smooth and resilient. Dehydration is present if skin is very wrinkled, withered, and/or dry. Edema may cause the skin to be taut and shiny. Bony prominences will be very pronounced on low body weight persons; obesity may result in extra folds where skin surfaces approximate causing friction and irritation to the skin.
Absorption and excretion	Observe skin surface for texture and moisture	Skin texture varies among individuals from dry to oily due to the amount of body secretions. Skin usually becomes more dry with advancing age; excessive perspiration may be due to environmental factors or elevated body temperature.
Protection	Observe for breaks in skin integrity	Normally there should be no eruptions of the skin surface.
Physical beauty	Observe texture, color, and general appearance of the skin	Blemishes, rashes, lesions, and discolored areas on the skin surface are warning signs of irritation or trauma.

Figure 3.4 Skin assessment factors. Observations not in accordance with normal findings indicate certain impairment of skin function that could increase pressure sore development risk. Temperature perception testing can be done using the Minnesota thermal disks of different thermal conductivities. The two-point discrimination test measures the minimal spatial discrimination ability of the patient on various parts of the skin (Gosnell 1987).

within 10 to 15 min of pressure removal. Persistence of the reddened area suggests that underlying tissue damage has already occurred.

A person carrying out the assessment should also be able to identify the stage of any pressure sores already present using the classification in figure 3.3 and recommend appropriate treatment if possible (Chapter 14).

3.2.2 Risk factor assessment

The assessment of pressure sore risk is complex and extensive because it involves many factors that exhibit much interaction and interdependence. Levine *et al* (1989a) classified risk factors into external factors which affect the skin through physical contact, and internal factors which are demographic and physiologic factors. External risk factors are pressure, shear force, friction, and moisture and internal factors are nutrition, activity level, bladder and bowel incontinence, and mental status. Braden and Bergstrom (1987) classify these factors according to their role as contributors to either of the two critical determinants of pressure sore development: (1) intensity and duration of pressure and (2) the tolerance of the skin and its supporting structure for pressure (figure 3.5). They further categorize the factors affecting tissue tolerance for pressure into extrinsic and intrinsic. Extrinsic factors influence tissue tolerance by impinging upon the skin surface. Intrinsic factors are those that influence the architecture and integrity of the skin's supporting structures and/or the vascular and lymphatic system that serves the skin and underlying

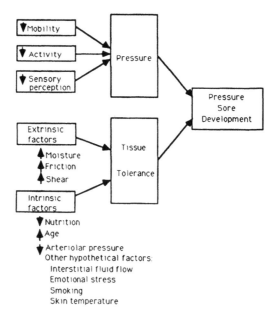

Figure 3.5 A conceptual schema for the study of the etiology of pressure sores which accounts for the relative contributions of the duration and intensity of pressure and the tissue tolerance for pressure. Arrows indicate the direction of change in the associated factor required to cause an increased risk (Braden and Bergstrom 1987).

structures. Copeland-Fields and Hoshiko (1989) polled the opinions of a group of 12 nurses in a rehabilitation department with respect to the relevance of the factors presented by Braden and Bergstrom (1987). They found that four of them were thought not to have significant relevance to pressure sore development (increased age, emotional stress, smoking, and skin temperature). The nurses also suggested many psychosocial factors such as patient lifestyle and knowledge.

In Section 1.5 we discussed the effects of pressure, shear force, friction, and moisture on pressure sore development and in Section 1.6 we presented several factors related to skin structure and function. Here we will discuss some of the other factors mentioned above. Braden and Bergstrom (1987) present an overview of pertinent literature. Figure 3.6 lists several established and hypothesized risk factors.

External	Internal	Psychosocial
Pressure	Mobility	Socioeconomic factors
Shear	Activity	Self-care
Friction	Mental status	Family dynamics
Moisture	Incontinence	Lifestyle
Skin temperature	Sensory perception	Health beliefs
	Age	Hygiene
	Nutrition	Patient knowledge
	Smoking	

Figure 3.6 There are no clear boundaries in the classification of pressure sore risk factors since they are highly interdependent and exhibit complicated relationships.

Mobility
A patient with diminished ability to change body position has an increased probability of developing a pressure sore since he will be exposed to prolonged and intense pressure. Spinal cord injured persons, especially those with injuries at high anatomical levels, have an increased incidence of pressure sores because of their reduced mobility. Richardson and Meyer (1981) found the incidence of pressure sores to be higher among quadriplegic persons than among paraplegic persons. This can be attributed to the fact that paraplegic persons enjoy greater mobility than quadriplegic persons (Braden and Bergstrom 1987). Mobility is affected by several factors, including mental status, neurological impairment, use of sedatives, presence of pain, and orthopedic injury (Levine 1989). Chapter 12 discusses behavior to relieve pressure.

Activity
Some patients, like paraplegic persons, can change and control their body positions but are still subject to prolonged pressure over some skin areas. This is the case with wheelchair-bound patients who can lift their bodies but don't do that often enough because of diminished sensation. Other conditions like arthritis and cancer can lead to confinement to bed or wheelchair causing limited activity. Reduced activity levels would probably also imply other factors that contribute to pressure sore formation such as increase in minute ventilation, cardiac output, and venous return (Braden and Bergstrom 1987).

Sensory perception
Patients with diminished ability to perceive or respond to discomfort by changing position or requesting assistance are also susceptible to pressure sore development. This group includes paraplegic persons who are able to shift position but do not perceive the signal to shift. It also includes patients with decreased consciousness who may perceive pain but cannot respond by changing position or requesting a change in position. Examples of patients that fall into this category are cerebral hemorrhage patients and those under prolonged anesthesia and excessive dosages of sedatives or tranquilizers.

Mental status
Some investigators feel that mental status itself should not be considered a risk factor for pressure sores. A diminished mental status indirectly places an individual at risk for injury to soft tissue since numerous medications, including sedatives, tranquilizers, and analgesics cause decreased levels of consciousness and alter the overall physical condition.

Incontinence
Bladder and bowel incontinence are a major cause of moisture, local skin irritation, and secondary infection. Section 1.6 discusses the effect of moisture on the skin's tolerance. Urea-splitting organisms produce ammonia which causes an alkaline burn especially if skin is sodden with moisture.

3.2.3 Assessment instruments

Pressure sore assessment instruments in general use ratings of risk factors in order to arrive at a number that reflects an individual's risk of developing a pressure sore. They are designed with the aim of providing a comprehensive, systematic, and efficient manner of gathering pertinent information. There are several such instruments that were and still are being clinically analyzed and evaluated. These assessment instruments may also include ones that focus on data regarding actual pressure sores after they have occurred. In this chapter however, we only consider the risk assessment instruments.

The best known assessment instrument is undoubtedly the Norton scale (Norton *et al* 1962). Figure 3.7 shows how Norton *et al* (1962) rated five risk factors on a scale of 1 to 4 each. A person carrying out the assessment decides, through examination of the physical state and the patient's record data, the rating of each risk factor and sums up all the ratings. A total score of 14 and below means that the person assessed is at high risk of developing a pressure sore and requires the institution of preventive measures. Norton *et al* (1962)

A Physical cond	B Mental state	C Activity	D Mobility	E Incontinence
Good 4	Alert 4	Ambulant 4	Full 4	None 4
Fair 3	Apathetic 3	Walk/help 3	Slight. limited 3	Occasional 3
Poor 2	Confused 2	Chairbound 2	Very limited 2	Usual/urine 2
Very bad 1	Stuporous 1	Bedfast 1	Immobile 1	Double 1

Figure 3.7 The Norton score uses five risk factors, each rated on a 1 to 4 scale. A total score of 14 or below indicates that the patient is at risk of developing a pressure sore (Norton *et al* 1962).

found a nearly linear relationship between the patient score and the incidence of pressure sores (figure 3.8).

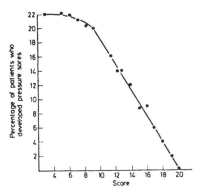

Figure 3.8 Results of a study done by Norton *et al* (1962) to determine the relationship between the score on the Norton scale and incidence of pressure sores. An almost linear relationship exists with deviation from linearity at low scores.

Goldstone and Goldstone (1982) conducted a study using a weighting scheme and discriminant analysis to evaluate the Norton scale. They concluded that it is a reliable guide to the incidence of pressure sores, although it does have a tendency to overpredict pressure sores where they do not in fact materialize. The study also suggests that physical condition and incontinence are the factors that perform best as discriminants and should therefore be correctly decided.

There are some drawbacks to the Norton scoring system. The vague terminology in the general physical condition category may allow for too much subjective interpretation. Further, both physical condition and mental status may predict mobility and activity and thereby contribute to duplication of categories. Nutritional status, in addition, is not assessed by the Norton scale (Levine *et al* 1989a). There have been several attempts to overcome these drawbacks. Gosnell (1973) developed a modified Norton scale in which she substituted the physical condition category with a nutrition category, rated the mental status category on a scale from 1 to 5 adding the state of unconsciousness, and included other factors such as skin appearance, tone, and sensation. The latter three factors were not rated numerically. Waterlow (1985) developed an instrument with the categories of: build/weight for height, visual skin type, continence, mobility, sex, age, and appetite. In addition, a special risk category includes: poor nutrition, sensory deprivation, smoking, high dose anti-inflammatory or steroids in use, and orthopedic surgery or fracture below the waist. These factors were rated nonuniformly. A total score of 10 or higher indicates an at risk status.

Using a somewhat independent approach, Bergstrom *et al* (1987) developed the Braden scale following the outline for the study of the etiology of pressure sores presented by Braden and Bergstrom (1987a) (Section 3.2.2). The Braden scale is composed of six subscales that reflect sensory perception, skin moisture, activity, mobility, friction and shear, and nutritional status. Each area has three to four levels with a one- or two-phrase description of qualifying attributes. Figure 3.9 shows the sensory perception subscale.

Braden Scale Risk Predictors For Skin Breakdown Patient's Name_____ Evaluator's Name _____. Date of assessment							
Sensory Perception Ability to respond to discomfort	*1.Completely limited:* Unresponsive to painful stimuli, either because of state of unconscious-ness or severe sensory impairment, which limits ability to feel pain over most of body surface.	*2.Very limited:* Responds only to painful stimuli (but not verbal commands) by opening eyes or flexing extremities. Cannot communicate discomfort verbally. OR Has a sensory impairment which limits the ability to feel pain or discomfort over 1/2 of body surface.	*3.Slightly limited:* Responds to verbal commands by opening eyes and obeying some commands, but cannot always communicate discomfort or need to be turned. OR Has some sensory impairment which limits ability to feel pain or discomfort in 1 or 2 extremities.	*4. No impairment:* Responds to verbal commands by obeying. Can communicate needs accurately. Has no sensory deficit which would limit ability to feel pain or discomfort.			

Figure 3.9 The sensory perception subscale of the Braden scale. Each grade has a clear description of qualifying attributes to make the assessment more accurate (Bergstrom *et al* 1987a).

The authors of the Braden scale leave the choice of the cut-off point determining risk status to the institution using the scale. They suggest that the determining factor be a balance between the cost of preventing and treating pressure sores with the decision being tempered by ethical principles.

Investigators conducted several studies to determine the reliability and validity of the Braden scale (Bergstrom *et al* 1987a, Bergstrom *et al* 1987b). The studies show that the Braden scale had a lower tendency to overpredict pressure sores than the Norton scale. In one study, the Norton scale overpredicted by 64% while the Braden scale overpredicted only by 36%.

Validity and reliability are the two basic requirements of a good pressure sore assessment instrument. Validity refers to the instrument's ability to truly identify those persons it claims to identify and it is measured by sensitivity and specificity. Sensitivity is the number of those who develop pressure sores and whose scores fall in the at-risk range of the instrument as a percentage of the total number of persons with pressure sores. Specificity, on the other hand, is the number of those who do not develop pressure sores and whose scores fall in the safe range of the instrument as a percentage of the total number of persons without pressure sores. The formulae for computation are:

$$\text{sensitivity} = \frac{\text{number of persons scoring ``at-risk'' and developing pressure sores}}{\text{total number of persons with pressure sores}} \times 100$$

$$\text{specificity} = \frac{\text{number of persons scoring ``safe'' and not developing pressure sores}}{\text{total number of persons without pressure sores}} \times 100$$

Reliability of an assessment tool refers to its ability to identify the same persons regardless of the user and it is evaluated by percent agreement or correlation measurements (Taylor *et al* 1988). Gosnell (1987) gives some criteria for choosing an assessment instrument that include the purpose, format, and content foci. Figure 3.10 lists some of the results of studies done to evaluate the sensitivities and specificities of assessment tools.

Tool/study	Sensitivity (%)	Specificity (%)
Norton/Norton *et al* 1962	63	70
Norton/Goldstone and Goldstone 1982	89	36
Gosnell/Gosnell 1973	50	73
Braden/Bergstrom *et al* 1987a	100	90
Braden/Bergstrom *et al* 1987b	83	64

Figure 3.10 Validity studies results. Taylor *et al* (1988) point out that caution should be used when comparing these figures because (1) different definitions of pressure sores could have been used, (2) scores used in calculations may have been acquired at different times, (3) different methods may have been used to obtain scores, and (4) other methodologic variables may have influenced the results.

3.3 OBJECTIVE EVALUATION OF SKIN STATUS

Early detection of pressure sores which relies upon inspection of skin color as discussed in Section 3.1 suffers from several difficulties: (1) a certain degree of subjectiveness is involved in the assessment such as in differentiating between persistent redness and normal short-term reactive hyperemia, (2) in the more advanced stages of pressure sore formation, the area may become ischemic with often only a margin of redness, (3) deep tissue damage may have already taken place while the surface of the skin shows only a small ulceration (closed pressure sores), (4) skin color changes are difficult to detect in patients with dark skins (Ferguson-Pell and Hagisawa 1987).

A more objective approach to skin assessment is to measure certain parametric changes in the skin that reflect the skin's viability. In comparison to techniques that measure pressure and other physical parameters at the body/support interface (Chapters 8 to 11), measurements of skin viability have the advantages of providing direct feedback of the amount of tissue damage, and bypassing the need to establish prescription criteria for physical parameters.

Successful objective detection of early warning signs requires high sensitivity in detecting the small changes in skin condition.

In this section we review some of the techniques that have been used to assess tissue viability objectively.

3.3.1 Thermography

Any body with temperature above absolute zero radiates electromagnetic power. The amount is determined by the body's temperature and physical properties. Bodies at room temperature (300 K) radiate energy predominantly in the infrared region of the spectrum. This principle is used in medical thermography to map the temperature of the human body on to a display surface. Thermographic scanners have been used successfully in several diagnostic applications such as early detection of breast cancer and determination of the depth of tissue destruction from frostbite and burns. Thermographs are usually presented on a cathode ray tube with gray levels corresponding to temperature.

Detection of the heat radiated from the body in the form of infrared radiation is accomplished by utilizing the principle of photoconductivity. A rotating mirror mechanical scanner images the object onto an indium antimonide detector cell cooled in liquid nitrogen (Siedband and Holden 1978).

Goller *et al* (1971) conducted a series of experiments using thermography on able-bodied subjects using a deadweight loading mechanism. They applied fixed pressure to the skin via Teflon applicators and acquired thermographs of the area at several times starting immediately after pressure removal. The results of their experiments suggested that there is a possibility of a correlation between the pressure applied to the skin and the temperature response of the skin immediately after removal of the pressure. Verhonick *et al* (1972) carried out a similar experiment but with typical body loading of the skin matching situations of bed rest. They observed a local increase in skin temperature after pressure removal reaching a maximum in approximately 3 min and then cooling back to normal. They also reported that the longer the pressure was applied, the longer was the time required by the skin to return to normal temperature. They suggested the utilization of this observation in determining a patient's susceptibility to pressure sore development by applying a specific pressure to the skin for a period of time and then noting the response time required by the skin to return to a normal temperature distribution. Trandel (1975) used a patient lift bed that allowed observation of the tissue subjected to body load in order to correlate visual and thermal responses. Their results showed that the thermal flare persists longer than the visible flare and attributed that to continued local elevated metabolic rate caused by the previous engorgement of blood. They reported a maximal increase in temperature during reactive hyperemia of 12°F. Pye and Bowker (1976) reported similar results.

Newman and Davis (1981) used thermography In a geriatric unit as a predictor of pressure sore formation by acquiring a thermogram of the buttocks. They compared predictions based on thermography alone to the Norton score and found that thermography provides a more precise indication of the risk of early development of a sacral pressure sore. Francis *et al* (1979) used a theoretical basis presented by Love and Lindsted (1975) to derive skin blood flow measurements from thermographic measurements.

3.3.2 Photoplethysmography

The basic principle of photoplethysmography is the modulation of reflection, transmission, and absorption properties of light as a result of volume changes. A light source and a photosensitive detector are arranged so that the detector can measure the intensity of light reflected from or transmitted through a capillary bed (figure 3.11). Blood has a light absorption coefficient that is higher than that of surrounding tissue, thus an increase in the amount of blood causes a corresponding decrease in the intensity of light detected. Photoplethysmography does not provide accurate volume measurements because the signal detected is very small. Furthermore, it is very sensitive to motion.

LEDs that produce narrow band spectra in the infrared range are preferred over tungsten lamps as light sources in photoplethysmography because they are less bulky and do not generate heat that may interfere with the measurement of blood flow. Photosensors are formed using silicon phototransistors.

Hertzman (1938) was the first to use photoplethysmography to estimate the blood supply of various areas of the skin. In spite of the simplicity of his device, his results demonstrated that photoplethysmography can be used to estimate the amount of blood flow in the skin.

Lee *et al* (1979) used photoplethysmography to assess the healing potentials

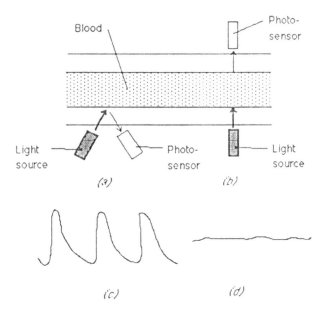

Figure 3.11 Light emitted from the source is partly absorbed by blood and tissue, the remaining part is reflected or transmitted and is detected by the photosensor; (a) reflection method, and (b) transmission method. (c) A tracing demonstrating good pulsatile cutaneous blood flow. (d) A relatively flat waveform indicates poor cutaneous blood flow.

of pressure sores. The motivation behind their work was to objectively determine the need for surgery to treat a pressure sore, thus avoiding the risk and cost of unnecessary surgery. The study used measurements of blood flow around the ulcer to determine the skin's potential for healing. Giltvedt *et al* (1984) developed a multiwavelength photoplethysmograph to detect simultaneous flow variations at levels of the dermis and deep vascular plexi. The principle of the multiwavelength device is that light of different wavelengths penetrates the skin to different depths. Thus two light sources emitting radiation at different wavelengths are reflected from blood vessels at different depths.

3.3.3 Oxygen measurement

Noninvasive transducers have been used to determine the effect of load on the partial pressure of oxygen in tissue measured at the skin surface (PO_2). Measurements of PO_2 under the sacrum, trochanter, lateral aspect of the thigh, and the ischial tuberosities showed a monotonic decrease with increasing applied pressure. Cutoff pressure was determined to be that at which the transducer read a zero value of PO_2 (Newson *et al* 1981, Newson and Rolfe 1982). Seiler and Stahelin (1979) investigated the effect of location on pressure–PO_2 relation. They loaded the skin both at bony prominences and at soft locations while measuring the skin surface PO_2. The results showed clearly that PO_2 drops much lower at bony prominences than elsewhere for the same amount of applied load. Seiler *et al* (1986) used these results to evaluate a certain bed rest position and mattress.

3.3.4 Biochemical measurements

Detection of biochemical changes in the skin tissue during the early stages of pressure sore formation can be used in a predicting procedure. However, such detection would normally require a certain amount of invasiveness and is clinically unacceptable. Ferguson-Pell and Hagisawa (1987) tried to use sweat from underneath loaded areas to provide this information. The reasoning behind this approach is that sweat glands are endowed with a capillary blood supply that may carry biochemicals signaling tissue distress. They collected sweat during pressure application on the forearm of a subject and found a significant increase in lactate level during ischemia but not afterwards. In a study conducted on pigs, Hagisawa *et al* (1988) reported a marked increase in serum creatine phosphokinase levels that persisted even after a week.

3.4 STUDY QUESTIONS

3.1   Does ischemic pallor only occur while pressure is applied or does it also occur after pressure release?
3.2   How long must pressure be applied to produce reactive hyperemia?
3.3   How long must reactive hyperemia last to indicate risk of pressure sores?
3.4   Describe objective measures of ischemic pallor and reactive hyperemia.

3.5   Which of the skin parts is involved in each stage of pressure sore development listed in figure 3.3.

3.6   Which pressure sore assessment scale is best for all patients? Explain.

3.7   Which pressure sore assessment scale is best for paraplegic patients? Explain.

3.8   A patient is in poor physical condition, has confused mental state, walks with help, mobility is slightly limited, and usually has urinary incontinence. Compute the probability of developing a pressure sore.

3.9   Explain how photoplethysmography can provide a measure of blood flow.

3.10 Explain how oxygen measurement can assess skin viability.

3.11 Explain how thermography can assess skin response to pressure.

4

Behavior to Prevent Pressure Sores

Basel Taha

Prevention of pressure sores is dependent on accurate assessment and appropriate action by either the person at risk or the person responsible for his care. In this chapter we present the behavior of different risk groups to combat the danger of pressure sore formation.

4.1 ABLE-BODIED PERSONS

Able-bodied persons having complete mobility and sensation are not subject to pressure sore formation since they perform weight reduction movements automatically during sitting and sleeping. The frequency of these movements is determined directly by the body's physiological needs to maintain a healthy state. Keane (1978-79) suggests that during sleep these requirements are the minimal necessary. Based on the observation that, during an average 8-h sleep on a soft bed, an able-bodied person performs one gross postural change every 11.6 min, Keane suggests that this is the minimal physiological mobility requirement. The only stable positions during sleep are the supine and the lateral with the legs flexed positions.

4.2 SPINAL CORD INJURED PERSONS

Spinal cord injured (SCI) persons spend most of their time in the wheelchair. Depending on the level of injury, they may have limited or no control of the trunk muscles. This lack of control has serious implications on the seating posture that lead to increased risk of pressure sore development. Kyphotic sitting is a term that refers to the seating posture assumed by wheelchair dependent persons due to this problem (figure 4.1). The "locking" of the posterior lumbar facets, referred to as lordosis, when sitting provides the able-bodied person with the ability to maintain a normal upright seating posture. However, due to the "unlocking" of these facets, SCI persons seated in an upright position are subject to forward loss of balance due to the anterior force

50

component of weight. As a way of counteracting this effect, a SCI person tends to assume a seating posture that leads to forward sliding of the buttock on the wheelchair seat and the settling of the back low on the backrest. As a result, there is an exposure of the ischial tuberosities to higher values of pressure and shear. Other problems associated with such seating posture are elevated disk pressures and compromised respiratory capacity.

Poor sitting posture is usually easy to notice by observation. Nevertheless, it is sometimes helpful to have an estimate of the amount of curvature in a person's spine such as in the case of treatment by remedial physiotherapy. Ferguson-Pell *et al* (1980) describe a device used to measure the spinal curvature consisting of metal probes passing through an array of holes in a vertical peg-board frame. The ends of the probes touching a canvas replacement for the wheelchair backrest outline the curvature of the spine while the person is assuming his habitual position.

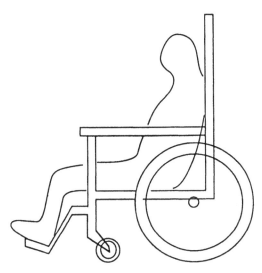

Figure 4.1 Because of lack of control of the posterior lumbar facets, a SCI person assumes a tilted posture that adds to the pressure under the ischial tuberosities.

One way to force the wheelchair dependent person to assume an upright position is to strap the chest to the backrest of the wheelchair. This however limits the range of reach for the person's hands and causes inconvenience for some persons like paraplegic persons. Using a lumbar support causes anterior tilting of the pelvis and thus a reduction of the ischial pressure. However, this approach is also subject to the problem of loss of balance mentioned earlier (Shield and Cook 1988). Zacharkow (1984) suggested reclining both the seat (10°) and the backrest (15°) to reduce the buttock sliding out of the chair. This would allow maximal use of a lumbar support and therefore reduce ischial pressure. Shield and Cook (1988) investigated the effect of the presence of a lumbar support in both the upright and reclined seat positions on the pressure under the ischial tuberosities. They measured the pressure under the buttocks of able-bodied volunteers using an ischiobarograph. Their results showed a significant reduction in the areas subjected to the highest pressures with the use of a lumbar support. This reduction was independent of whether the chair was upright or reclined which implies that the convenience of the inclination does not add to the problem of elevated ischial pressure.

Other factors that influence the seating posture of a wheelchair person are foot rest height and hamstring muscle tension. SCI persons sometimes have extremely tight hamstring muscles that may prevent the pelvis from anteriorly tilting with the use of a lumbar support. The lumbar support in this case only adds to the forward force pushing the person's buttocks forward.

In addition to considerations of seating posture arrangements to reduce pressure, SCI persons who are able to change body position by lifting up or by weight shifting are instructed to do so periodically (section 12.2). Those persons who cannot perform such action should be assisted to move or leave the wheelchair regularly.

4.3 AGED PERSONS

Chapters 1 and 3 pointed out that old age plays an important role in the increased susceptibility to pressure sore development. An especially high-risk group are geriatric patients. Many elderly patients in geriatric wards are confined to a bed most of the time, putting them in high danger of developing pressure sores. Mobility is affected by several factors that are not uncommon with old age patients such as reduced mental status, neurological impairment, use of sedatives, presence of pain, and orthopedic injury (Levine *et al* 1989a).

Studies of mobility of elderly patients in bed during sleep (Barbenel *et al* 1986, Nicholson *et al* 1988) have shown that, in general, the number and frequency of movements made by the elderly patients were small. Furthermore, patients that scored high on the Norton scale (high-risk) made fewer movements than those who scored low (low-risk). Provided that a patient's mobility is not limited such as by respirator dependence, pain, obesity or orthopedic injury, turning at 2-h intervals is recommended. Draw sheets can be used to reduce the effect of shear during positioning of the patient.

Besides reduced mobility, geriatric patients are also prone to have bowel and/or bladder incontinence which in addition to the risk associated with moisture brings in the dangers of skin irritation and infection. It is therefore recommended that skin be washed during normal baths and at other times if there has been incontinence. Too frequent washing should be avoided since it removes the skin's surface liquids. In addition, alcohol and methylated spirits cause vasoconstriction in already ischemic areas and may increase the possibility of developing a pressure sore. Barrier substances such as oil, silicone or zinc base may be useful for patients who are incontinent and require frequent washing. A better way to avoid the dangers of urinary incontinence is through the use of external catheters for males or urinary collection devices for women. Fecal incontinence collectors can be used for patients with bowel incontinence. Toileting schedules is still another alternative. These alternatives should be weighed according to the cost and staff availability.

4.4 OPERATING TABLE

Patients at risk of pressure sores undergoing surgical procedures require special interventions. Support surfaces designed to relieve pressure may not be appropriate for use in an operating room for reasons of stability, safety, and realistic considerations for the operating room environment. For example, a water or gel mattress will leak if accidentally punctured with a needle, which could mean a disaster in the operating room. Air-filled mattresses are unstable and cause the patient to move. In addition, performing an emergency cardiopulmonary resuscitation necessitates the use of a firm surface for adequate compression of the chest. Due to these special considerations, each patient's case should be considered separately and appropriate action taken with regard to available resources. Arnell (1988) points out the shortage of research on good, effective, support surfaces for use on the operating table.

4.5 PRESSURE MANAGEMENT TEAM

Prevention of pressure sores in the hospital is a complicated process involving the assessment of complicated factors and accordingly the recommendation of interventional strategies. The traditional idea that the nurse alone should be responsible for the prevention of pressure sores is therefore ineffective. A team of nursing and medical staff responsible for this task can be most effective in the prevention of pressure sores. Nevertheless, the nurse remains the most essential part of such a team and acts as the main source of information about the patient condition. Levine *et al* (1989a) suggest that such a team for geriatric care in particular include a nurse, nurse's aide, rehabilitation nurse, dietitian, physical therapist, occupational therapist and preferably an enterostomal therapist. With the physician acting as a coordinator and each member of the team interacting with other members, such an approach leads to improved diagnosis, suitable device prescription and, of course, reduced pain and cost to the patient through prevention. A quarterly periodical is available to assist the professional management team in managing, prevention, and treatment of pressure sores (*Decubitus: the journal of skin ulcers*, 103 North Second Street, West Dundee, IL 60118).

For the SCI person, where the pressure sores are due to seating, the concept of a pressure clinic is a more appropriate approach. The pressure clinic is a special clinic for the SCI outpatients and newly injured inpatients. Its main objectives are: (1) To educate the patients about the problem of pressure sores and advise them about action to prevent their occurrence. (2) To assess each patient's risk of developing pressure sores considering psychological and social aspects. (3) To establish a link with the patient after discharge to ensure appropriate action and validity of any prescribed devices. (4) To reinforce patient education in pressure prevention. (5) To collect data for future research. Such a clinic would optimally be staffed by a nurse, a therapy assistant, physiotherapist, and occupational therapist and a community liaison nurse. In a recent evaluation of such a clinic in a spinal injuries unit, Rothery (1989) reports that only 4% of the patients who completed their rehabilitation in the center developed a pressure sore after discharge.

4.6 PATIENT COMPLIANCE OUTSIDE INSTITUTION

Parish *et al* (1983) summarize the subjects that should be covered in counseling, films and literature distributed to patients discharged to home care:

1 "An urgent explanation of the nature, causes and dangers of pressure sores and the urgent need to prevent them.
2 Instruction in the establishment of a regular position shift schedule to relieve pressure.
3 An outline of the danger signs of impending pressure sores together with details for the establishment of a systematic skin inspection program, including the use of mirrors, to spot the signs in their earliest stage of development.
4 A detailed list of the everyday environmental hazards that can damage skin and lead to the development of sores—foreign materials in the bed, heating pads, wrinkled linen, careless use of the toilet, etc.
5 The development of bowel and bladder programs to prevent or minimize contact of urine and feces with the skin.
6 A dietary regimen that will provide adequate protein to maintain a positive nitrogen balance, sufficient calories to maintain body weight and minimum daily requirements of vitamins and minerals.
7 A program for dealing with small and early pressure sores that includes instructions on suitable topical therapy, bridging with pillows, and proper hygienic precautions.
8 Reminders every step of the way that the responsibility for the preventing pressure sores in home settings rests with the patient *himself*, even when he is fortunate enough to have attendants to assist."

A patient's compliance with instructions about pressure sore prevention techniques is dependent upon several psychological, social and economic factors. In general, investigators have stressed the importance of self-care, self-responsibility and level of motivation in preventing pressure sores. These factors are greatly affected by the patient's psychology and general feeling about his condition. Family care is another important factor.

Krouskop *et al* (1983) report the results of the introduction of psychological counseling and an active education program to promote self-awareness and consciousness of pressure damage on the recurrence of pressure sores in patients. Prior to this program 9% of the patients treated at the clinic and discharged returned with repeated tissue breakdown. After the introduction of the program this number dropped to only 4% in one year.

4.7 STUDY QUESTIONS

4.1   How does the average interval between gross postural changes made during sleep vary with compliance of the bed surface?

4.2   How should the average interval between pressure reliefs for paraplegic persons vary with compliance of the seat cushion?

4.3   How should the average interval between turnings vary with compliance of the bed surface?

4.4   If recommendations for cleanliness are not followed, how can this cause pressure sores? Consider effects of perspiration, urea, feces, and removal of oils in skin due to excessive cleaning.

4.5   How can a dietician assist in a pressure management team?

4.6   How can an occupational therapist assist in a pressure management team?

4.7   What factors affect the recurrence of pressure sores for patients outside the institution?

4.8   How can a rehabilitation engineer assist the pressure sore prevention team?

4.9   How well do patients comply with pressure sore prevention recommendations outside the institution?

5

Seat Cushions

Steven Tang

Seat cushions have been used commercially to promote better comfort in furniture designs and in travel vehicles. In this chapter, we examine the use of seat cushions as preventive devices against pressure sores. Seat cushions are currently the most popular and highly researched method for pressure sore prevention of the spinal cord injured. The basic goal of cushions is to reduce and redistribute interface pressures for the patient by supplying a compliant support at the patient–seat interface. We discuss several common types of materials used for cushion fabrication, as well as various means of cushion design. Finally, we describe objective and subjective criteria for evaluating the performance of seat cushions. Ferguson-Pell (1990) presents an extended review of seat cushion selection and notes that no single cushion meets the needs of all users.

5.1 CUSHION TYPES AND MATERIALS

Researchers have tried several different material types for seat cushion design. Most materials are chosen for their ability to ultimately reduce pressures near bony prominences using simple principles of elasticity and Pascal's law. Every material has particular characteristics that create different patient seating environments. There are numerous advantages and disadvantages to each type.

5.1.1 Foam cushions

The most common type cushion is made from medium density polymeric foam. Polymeric foams come in the different cell structures shown in figure 5.1. There are three basic types of cell formation: closed, open, and reticulated (Meinecke and Clark 1973). Closed cell foams consist of polyhedron structures enclosed by membranes of variable thickness. Closed cell foams are generally more rigid in compliance. An open cell foam has interconnected, perforated membranes, allowing greater flexibility and better ventilation. Most foams have a mixture of open and closed cell structures. They are characterized as fraction open cells (Baer 1964). Reticulated materials have a larger pore size and contain thinner membranes that have been chemically or heat treated. Reticulated foam is less durable than open cell foam but has a greater elastic performance. When manufacturing foam, the individual cell size and geometry can also be varied to provide different structural characteristics.

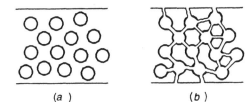 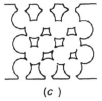

(a)   (b)   (c)

Figure 5.1 Polymeric foams used for cushions. (a) Closed cell. (b) Open cell. (c) Reticulated cell.

A commonly used polymer in foam cushions is medium density polyurethane. The Rogers 1836 is a standard elastic foam cushion. Polyurethane foams are noted for their lightweight and flexible properties. They are used to create resilient, easily transferrable cushions. Ideally they exhibit very linear characteristics of stress σ versus strain ε :

$$\sigma = E\varepsilon$$

where E is Young's modulus. However load–indentation curves exhibit nonlinear relationships.

These foams are less expensive than most other materials because they are readily available in bulk quantities. Higher density polyurethane foams can be carved into specific contours that conform to the patient's buttocks. The main drawback to polyurethane foams is that they do not endure much environmental aging (Noble *et al* 1983). Most foams last about six months before losing a significant amount of elasticity. Protective covers are necessary to extend the lifetime of foam cushions. Rulings in California and Boston by fire officials have restricted the use of traditional foams for wheelchair cushions for safety purposes (Krouskop *et al* 1986a).

Viscoelastic foams (also called "memory" foams) are very similar to regular polyurethane foams with the exception of having a higher damping factor. They can provide improved stability as they dampen loads rather than cause the patient to bounce. They exhibit time dependent stress–strain characteristics:

$$\sigma = \eta \frac{d\varepsilon}{dt} .$$

This is important for paraplegics who require a firm "reaction point" for self-propulsion in a wheelchair. The mechanical properties of viscoelastic foam can be modeled by the spring–dashpot system in figure 5.2. A viscoelastic material can be characterized as the combination of an ideal elastic solid and a perfectly viscous fluid. The resulting material is one with a nonlinear, time-dependent stress–strain relationship. The AliMed 164-S is one type of viscoelastic cushion.

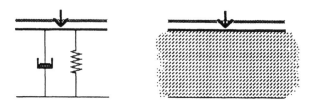

Figure 5.2 Viscoelastic foam can be modeled as spring–dashpot system.

The VASIO cushion (Veteran's Administration Seating Interface Orthosis) is an example of a hybrid cushion. It is made from two-part polyurethane foam with an insert of Ethafoam-220, a more rigid type polymer foam (Perkash *et al* 1984). The Ethafoam is positioned beneath the proximal thigh to provide better support and balance (figure 5.3). The insert acts as a fulcrum to transfer pressures from the pelvis toward the legs and thighs. A hybrid cushion makes use of both rigid and elastic polymer properties to provide stability without compromising compliance quality.

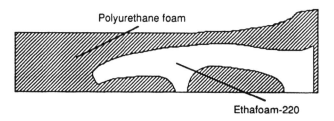

Polyurethane foam

Ethafoam-220

Figure 5.3 This is a lateral cross-section of the VASIO-P. The cushion has a rigid Ethafoam insert to transfer pressures from the pelvis toward the legs and thighs (Perkash *et al* 1984).

5.1.2 Air-filled cushions

Air-filled cushions (AFCs) attempt to reduce pressure under the ischial tuberosities by evenly distributing the pressure across the whole body area at the cushion interface. The motivation for such a cushion comes from the application of Pascal's law which states that pressure exerted at any point upon a confined liquid is transmitted undiminished in all directions. AFCs are advantageous because of their lightweight composition and longevity. Some commonly used air cushions are the Roho, Bye Bye Decubiti, and Gaymar Sofcare.

The Roho has been the most researched AFC in recent years and has had mixed reviews; yet, it still remains the forerunner of all air flotation designs, and has become almost synonymous with the AFC class. Figure 5.4 shows a Roho cushion alone in its uncompressed state. The Roho now comes in various different schemes in the attempt to offer posture balance and customized specifications. The standard "High Profile" Roho has an array of 4-inch (100-mm) air bag cells that are interconnected at the base of the cushion. When the patient sits on the cushion, it deforms to the contours of the patient's buttocks. A pressure pump is supplied with instructions on how to properly inflate the cushion. While AFCs are not prone to environmental aging, they are susceptible to punctures and tears.

Figure 5.4 The Roho cushion has individual air cells interconnected at the base.

5.1.3 Gel or fluid floatation cushions

The gel or fluid cushion operates very similarly to an AFC but it has more effective damping abilities. A gel cushion can be thought of as a fluid analogy to the viscoelastic foam cushion except that it doesn't have nearly as much elasticity. These cushions tend to be much heavier than air flotation devices and also more expensive (Conine *et al* 1989). Because of this, they are quite cumbersome and unstable during patient transfers. The Stryker flotation pad is one type of gel cushion. More common are cushions that have a smaller gel pouch on top of a contoured polyurethane foam base. Gel cushions are noted for their capacity to conduct heat away from the patient to avoid increased temperatures at the skin interface.

5.1.4 Particle-filled cushions

The particle-filled cushion (PFC) is a bean bag type cushion and is also generalized as the bucket seat cushion. A typical PFC is made from an airtight flexible bag containing small polystyrene beads. Using a vacuum as a pump, the bag is inflated to provide space between the particles. After the patient sits in the bag, air is removed so that a tightly contoured mold remains around the patient. Francis Jones and Evans (1978) recommended PFCs for a particular minority patient group: spastic paraplegia and tetraplegia, cerebral palsies, patients with asymmetric shape due to spinal and hip deformities and those with abnormal posture reflexes. They cited the need of these patients for firmer support at the pelvic region because of the spastic behaviors and tendencies to fall out of shallow cushions. Francis Jones and Evans believed that PFCs are not suitable or profitable for active and independent paraplegic persons who are able to do their own transfers. Hobson *et al* (1986) discussed a more permanent bead matrix system by mixing in an adhesive resin before the final shaping and evacuation of the air. The resiliency of the resin–bead matrix enhances impact strength and provides a degree of cushioning.

5.2 METHODS OF DESIGN

Much research has been devoted toward the design of seat cushions in the attempt to improve old methods of construction. Techniques range from totally prefabricated patterns to individually custom contoured cushions. Computer-aided methods have also been used to facilitate and expedite the process.

5.2.1 Precontoured support structures

Most of the foam cushions currently available are contoured in some fashion, although clinics still utilize flat slabs for simple padding purposes. Precontoured cushions were developed with the hopes of creating a universal cushion to fulfill the "one size fits all" idea. Lim *et al* (1988) evaluated a simple cut-out scheme (figure 5.5) that could be used for quick mass production. The basic design began with 25 mm of a medium-density polyurethane foam glued on top of 50 mm firm chipped foam. A preischial bar 38 mm thick made from polyurethane foam was glued on top and a cut out was made at the ischial tuberosity (IT) region. The cut-out length was based on the IT indentation measurements of a patient sitting on a 50 mm Tempra-foam cushion. The purpose of the cut-out is to alleviate pressures near the sacrum by having the buttocks hang in the cut-out area.

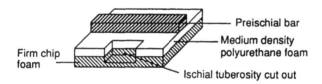

Figure 5.5 The precontoured foam cushion design provides an ischial tuberosity cut-out (Lim *et al* 1988).

Out of a patient group of 52 elderly subjects, Lim *et al* did not find any statistical differences in the incidence, location, severity, or healing time of the pressure sores that developed between subjects who used the slab and contoured form. The tests were conducted over a 5-month period. However, of the sores which did develop among the slab cushion group, the more severe ones were found in the IT area. There were no attempts at comparing differences of IT pressures or pressure distributions between the two groups.

There have been two other types of cushion designs that followed the idea of an ischial cut-out region. One is the "Paracare" cushion which was developed by Key *et al* (1978) at the Conradie Pressure Clinic. The cushion is constructed from expanded polyurethane elastic foam of high density to prevent "bottoming out" (allowing the buttocks to reach a depth where there is no more elasticity in the foam). The basic design is similar to that in figure 5.5 except there is no preischial bar. A cut-out under the IT is made after measuring the lateral separation of the patient's ITs. Modifications are then made to create interface pressures below the following maxima: 20 mm Hg under the IT, 50 mm Hg under the trochanteric shelf, 60 mm Hg under each preischial area, and 80 mm Hg under the anterior half of the thighs. The front portion of the cushion is also slightly raised with pads to further ensure relief at the ischii. The Paracare also has an optional backrest designed similarly to the seat cushion with a lumbar cut-out. In tandem, there is essentially a no-load condition over the sacrum, coccyx, and vertebral spinous processes.

In their most recent follow-up study, Key *et al* claimed to have an 89% success rate in preventing and relieving pressure sores. Specifically, 49 out of 56 spinal cord injured patients began with open sores of varying severity and had them heal while sitting on the cushion, and had no recurrence. Key *et al* did not attempt to record pressure distributions at the interface.

The other cushion design that works on similar principles is the VASIO cushion. It incorporates a hybrid style contoured foam base with a perineal recess to better control humidity and heat retention. Perkash *et al* (1984) developed the VASIO cushion in an attempt to reduce forces around the IT and promote proper spinal–pelvic alignment in seating posture. The design was constructed to minimize localized areas of high pressure and to redistribute pressure over the lateral gluteal muscles, subtrochanteric shelf, and the posterior thighs. Perkash *et al* considered pressures of 65, 70, and 80 mm Hg to be acceptable under the ischii, trochanters, and posterior thighs, respectively. These pressures are based on the study of Reswick and Rogers (1976) which indicated that the outer regions of the buttocks can sustain greater loads than the ITs.

The contours of the cushion were refined through eight iterations, with each modification attempting to move the interface pressures closer to the accepted ranges. Figure 5.6 shows the mean pressure distributions of the final cushion model (P refers to the "Paraplegic" model). It is clear that very little pressure remains in the sacral area.

Perkash *et al* cited a 52 out of 66 success rate. Success was defined as the number of patients who had interface pressures below the recommended maximum while sitting on the cushion. The cushion, however, seemed to be more effective for patients without skeletal deformities. This might be a result of the posturally restrictive nature of the VASIO cushion design.

5.2.2 Modular Design

Cochran and Palmieri (1981) developed another method of design. They suggested that there is currently no single material that has all the ideal characteristics for a preventive seat cushion. Materials with admirable features in one area often have deficient ones in another. Rather than researching new exotic materials, Cochran and Palmieri proposed combining different materials to form a multilayered cushion. In this way, the separate layers act together to enhance advantageous properties and offset the disadvantages of the individual layers.

Ferguson-Pell *et al* (1986) studied in depth the effectiveness of this modular system. They classified all the materials they chose to use into three categories:

foams, Temperfoams (viscoelastic), and gels. The generic construction was composed of a cover, three distinct layers, and a base. The cover segment was to be smooth, strong, light, flexible, and easily cleaned. The top layer would be of an open cell or reticulated foam for the purposes of encouraging air circulation and reducing sitting pressure. The middle layer would be of various medium- or high-density foams or a gel to reduce pressure and shear forces, and control heat. The bottom layer would be constructed from some type of medium viscoelastic foam to improve conformance and reduce pressures. Finally, the base would be made from a firm expanded foam to prevent the sag caused by the sling seat.

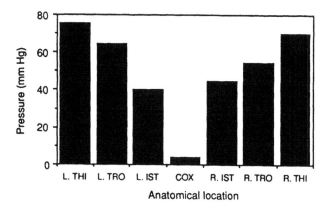

Figure 5.6 The mean pressure distribution on a VASIO-P cushion recorded by Perkash *et al* (1984). Anatomical locations: THI- Thigh, TRO- Trochanter, IST- Ischial tuberosity, COX- Coccyx.

Ferguson-Pell *et al* selected 13 different materials to test their design. The components ranged from low density foams to hard Temperfoams and gels. They created 32 prototype cushions from various combinations of the original materials. After testing the cushions mechanically using load-indentation, they narrowed the selection down to 13 particular cushion prototypes. Testing was done using able-bodied subjects to measure the interface pressures. Figure 5.7 lists the results of the test, showing the combinations that gave the lowest interface pressures. There are two particular combinations that appear across all weight groups and might be considered "good general cushion designs." One is the 2" (50 mm) medium foam/2" (50 mm) hard foam cushion, and the other is the 2" (50 mm) gel/1" (25 mm) medium temperfoam cushion. They measured an average IT pressure of 69 and 68 mm Hg for all three weight groups. These two offer a choice between an inexpensive foam combination and a heat-dissipating, shear-relieving, gel-foam cushion. Ferguson-Pell *et al* still recommended that special cushions not be interpreted as a totally preventive device for pressure sores and that frequent replacement of components is necessary to maintain proper function.

Weight group 1 (59–68 kg)	Weight group 2 (68–77 kg)	Weight group 3 (77–86 kg)
2" med foam/2" hard foam	2" med foam/2" hard foam	2" med foam/2" hard foam
2" med foam/1" hard foam	2" med T./1" very hard foam	2" gel/1" med T.
2" gel/1" med T.	2" gel/1" med T.	
1" gel/2" med foam	1" gel/2" med foam	
1" soft T./1" med T/1" very hard foam	1" soft T./1" med T./1" very hard foam	
2" med T./1" very hard foam		
1" soft T./2" med foam		

Figure 5.7 Cushion prototypes that gave the lowest interface pressures for different body weights. Materials are listed in order of their respective position from the top layer. With the exception of the 2" med foam/2" hard foam cushion, all prototypes are intended to have a 1" reticulated layer on top and a solid Ethafoam filler below. T = Temperfoam. (Ferguson-Pell et al 1986).

5.2.3 Custom contoured design

Numerous studies (Sprigle et al (1988), Sprigle and Chung (1989), Chung et al (1987), Sprigle et al (1990)) indicate that there is pressure reduction and better comfort from custom contoured cushions. Contouring increases the seating contact area and displaces pressure more uniformly than slab foams. Sprigle and Chung (1989) did tests on 11 subjects to see if there were significant pressure changes in using contoured cushions. Average pressures using the contoured foam cushions were 47 mm Hg versus 57 mm Hg for flat foam. Sprigle et al (1988) indicated that their test subjects felt stable and had a good sense of balance when using custom contoured designs. Figure 5.8 shows the pressure redistribution on a contoured polyurethane foam in comparison to the original slab. The contoured distribution has lower peak pressures and fewer pressure gradients. The main concern with the use of contoured cushions is that the patient must be more precise in finding the exact seating location. This is problematic since many wheelchair patients are spinal cord injured and have no sensation in the buttocks.

```
              25              33                        25              29
              29  Anterior    34                        24  Anterior    30
              19              33  Thighs                 26              35  Thighs
              27              28                          21              20
              23              30                          31              28
              30              45                          25              28

         41 37 17    81 52 28                       40 47 19    52 29 32
Buttocks 30 37 42    84 57 16              Buttocks  32 54 36    47 45 39
         69 44 75    42 45 42                        30 37 44    52 44 33
         32 34 43    41 32 33                        18 37 49    49 25 37

              Posterior                                   Posterior

                (a )                                       (b )
```

Figure 5.8 Pressures in mm Hg measured from sitting on a 3" (75 mm) slab of low density polyurethane cushion (a) and an equivalent contoured cushion (b) using an Oxford Pressure Monitor (Chung et al 1987).

There are several different ways of developing and manufacturing such a cushion. The simplest way is to create a foam contour from a plaster type mold of the patient (McGovern *et al* 1988). McGovern *et al*, in their design, first completed a one-piece mold of the patient and then used a manual cast measuring contour gage system to measure discrete levels of the contour (figure 5.9(*a*)). The gage was made from 168 parallel 1/8" diameter × 12" long (4 × 300 mm) stainless steel rods held by two aluminum channels. The rods could move independently from one another. Once the positional depths of the rods were set, the rods were removed from the cast and placed on a 1/2" or 1" (13 or 25 mm) thick Ethafoam blank slab. The contour was then cut out of the blank slab using a bandsaw. This process was repeated until the whole contour was represented by cross-sectional pieces of foam (figure 5.9(*b*)). The Ethafoam pieces were held together with three lengths of threaded rods pierced through the sections. Finally, the edges were smoothed with a rotating wire brush and a 1/2" to 1" (13 to 25 mm) polyurethane foam lining was placed into the support. For practical purposes, this design can be a very time consuming process.

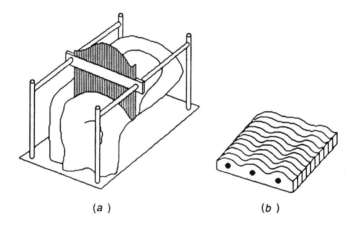

(*a*)   (*b*)

Figure 5.9 Custom contoured cushion design. (*a*) Cast measuring system using stainless steel rods. (*b*) Ethafoam slabs stack together to form the contoured cushion after being cut by a band saw (McGovern *et al* 1988).

Another method for constructing contoured foam cushions was developed by Hassard *et al* (1971). It initially followed the bead matrix cushion design by having the patient create a "female" impression mold. This was done with styrofoam beads enclosed in a plastic envelope attached to a vacuum pump. A solid impression would be left after the air was completely evacuated. Subsequently, a "male" mold was made by filling the "female" mold with a wet plaster. Once the plaster dried, a Plastazote liner was formed over the mold and coated with petroleum jelly parting agent. Finally, a two-part polyurethane foam mixture (Pedilen Rigid Foam) was poured into a shell with the "male" mold fitted on top. The hardened foam was then trimmed to fit the size of the wheelchair. A trial sitting was necessary to examine actual effectiveness. To check for localized pressures, a thin sheet of segmented clay putty was placed between two pieces of clear plastic. Areas of excessive pressure were revealed by flattened sections of clay strips.

Hobson and Tooms (1981) researched a method that bypassed many of the molding steps in the previous process. Essentially, this technique (called "foam-in-place") used the patient's own body as the molding device. A box-shaped mold was fabricated from urethane plastic and was closed on all sides except one. A sheet of thin latex (5 mm) was lightly stretched over the top. Once the patient was properly seated in the foaming chair on top of the latex, a two-part polyurethane foam (CPR 1947 N) was injected into the front of the mold. The reaction between the two foam components caused the foam to rise up and force the latex skin around the patient's buttocks. Within several minutes, the foam gelled into a soft cushion. There was a slight safety concern in applying this technique because there were toxic fumes that could be inhaled and toxic agents that could come in contact with the patient's skin. However, Hobson and Tooms concluded that the toxicity of the process was well within reasonable specified limits.

5.2.4 Ultrasound mapping

Aside from mechanical methods, there are also electronic methods of obtaining buttock–cushion interface contours to design cushions. One is to use ultrasound to measure the depth of indentation caused by a patient sitting on a typical polyvinyl chloride (PVC) gel material. Kadaba *et al* (1984) introduced a prototype ultrasonic system to measure contours using the principle of refraction. A transducer sends an excitation pulse that travels through the cushion medium (figure 5.10). The signal is reflected and refracted at each interface between two layers with dissimilar acoustic impedances. Thus, it is possible to accurately determine the thickness of the cushion based on the delay of the reflected echoes of the original pulse. As the transducer is positioned across the frame at marked intervals, the delay time of the echo is digitized and stored on a PC. This technique is not only useful for measuring buttock–cushion contours, but it might help to describe the geometry of soft tissues deformed under a load.

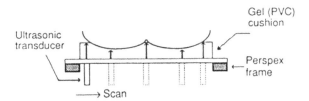

Figure 5.10 The ultrasonic transducer emits a pulse that measures distance to the buttocks by time of flight (Kadaba *et al* 1984).

5.2.5 Computer facilitated design

Neth *et al* (1989) enhanced the previous mechanical method of cutting Ethafoam blanks from a cast molding system. Rather than using long rods to mark the slabs for proper cutting, electronic linear potentiometers were mounted onto aluminum shafts and directly recorded a certain voltage for each section of the contour. Potentiometers were spaced at 1/2" (13 mm) intervals. Voltage was then sampled and sent to an analog-to-digital converter which was recorded by an IBM PC/AT. The graphics software can be used to drive a numerically controlled cutter which consists of a two-axis positioning table and a fixed cutting tool. This system can diminish fabrication time of custom cushions from weeks to hours.

This idea can be extended to a tri-axis milling machine to further simplify the manufacturing process. Brienza *et al* (1988) researched a computer-aided system (figure 5.11) that accurately cut a contoured shape straight from a flat foam cushion. This method does not necessitate the use of many foam slabs. It does, however, require use of higher density foams such as Ethafoam, or viscoelastic foam. Most commercially available foams can be carved through the milling machine. The system is programmed and run from a PC. The program drives a three-axis servomechanism and utilizes a velocity feedback control system. The goal of the system is to minimize time in producing a smooth contoured surface. There are two parameters that affect these desired goals: the tool size, and the cutting increment. Using a large tool speeds up the process but doesn't give as smooth a contour. Employing smaller cutting increments increases the resolution of the contour but is more time consuming. While the program runs, the computer numeric controller (CNC) linearly interpolates between the data points stored in the computer. This information is fed to the motor that cuts the foam surface. The tachometer measures velocity from the motor, and the optical encoder reads positional information. The CNC receives feedback from these two devices and readjusts the motor by sending an error signal to the pulse-width modulator. Figure 5.12 illustrates the cutting process using a spherical wire brush. Other types of tools can be used: double edged flat blades, router bits, or solid spherical cutters. Brienza *et al* claimed the total time required for patient evaluation, contour measurement, and cushion manufacturing to be about one hour.

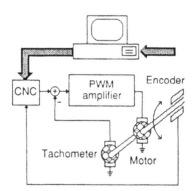

Figure 5.11 The cushion cutting system block diagram. Contour data are entered into the computer. Computer numeric controller (CNC) software operates a three-axis milling machine via a pulse width modulator (PWM) amplifier. The optical encoder and tachometer provide feedback to the CNC (Brienza *et al* 1988).

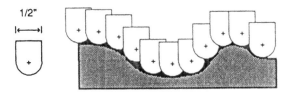

Figure 5.12 A spherical wire brush spins at 4200 rpm to cut the cushion (Brienza *et al* 1988).

Computer systems not only diminish the time spent in actual cushion design, but they also help assist in the decision-making process for prescribing wheelchair cushions to individual patients. Trachtman *et al* (1984) developed software that attempted to best fit a patient to one of the commercially available cushions. Their project goal was to organize the cushion-fitting procedure so that there is a direct path to a particular decision. The program requires initial data about the patient's administrative, clinical, wheelchair, and skin condition. Next, the patient is directed to have pressure measurements taken while sitting on some reference foam cushions. The algorithm then processes this information and determines set goals for the patient, such as maximal allowable pressures, stability, temperature, and comfort criteria. Using a cushion summary chart, which contains an updated record of all cushions evaluated by the program, the algorithm provides the evaluator with different options. A continual matching process occurs between goals and possible solutions until the optimal cushion is selected. Until a universal cushion is found, this type of software can be useful in making selections from a diversely growing number of seat cushions.

5.3 CUSHION INTERFACE CONDITIONS

Creating a viable cushion requires finding the right materials and utilizing a proper design to reduce pressures. It also involves looking at the many interface conditions that arise in normal use of such a cushion. Taking a clinical cushion and making it suitable for everyday use requires investigation of how the environment affects the cushion.

5.3.1 Environmental protection for cushions

Although repetitive sitting on foam cushions has been cited as a main factor for material fatigue, there is also evidence that pure environmental aging from outdoor exposure is equally damaging. Noble *et al* (1983) studied the direct effect of the environment on polyurethane foams by placing cushions (with and without covers) of different densities and hardness in an outdoor setting for a period of six months. Immediate observations of the aged foams showed discoloration and crumbling of the surface layers. Uncovered test pieces displayed an average loss in thickness of 7%. In all cases, the foams were found to harden significantly over the first 10 days and then continually decrease in hardness to about –40% over a 3–6 month period.

Foams which were hardest initially were found to eventually soften the most, while softer foams hardened more dramatically during the initial states of aging. Hardness was measured by examining the amount of force necessary to indent the foam a certain fraction of its original thickness. Oxidation seemed to be a possible factor in the cause of this problem. Figure 5.13 shows a time plot of change in hardness in foams. It is apparent that there are few times in the aging of a cushion where the hardness is close to its initial value. The conclusion by Noble *et al* was that the impact of aging can be greatly reduced by selecting foams of the highest practicable density in combination with relatively low hardness. An alternative approach would also be to environmentally pre-age foam cushions before use.

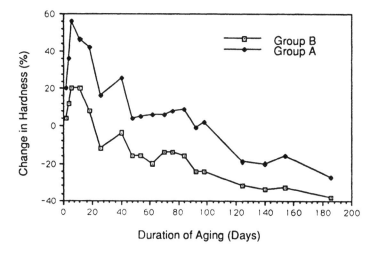

Duration of Aging (Days)

Figure 5.13 Foam hardness vs. time. Group A foams have a mean hardness of 43 lbs (191 N) (at 25% Indentation Load Deformation). Group B foams have a mean hardness of 58 lbs (236 N) (Noble *et al* 1983).

In addition to environmental aging, polyurethane foams are subject to stiffness due to water content. This can be an important factor for patients who are incontinent or patients who come in contact with liquids during meals. Brown and Pearcy (1986) evaluated the effect of water content on the compressive characteristics of foam cushions. Tests were performed on two grades of polyurethane foams: a soft yellow foam with a density of 49 kg/m^3 and a chip foam conglomerate with a density of 99 kg/m^3. The soft foam had a 0.93 void to solid ratio and a permeability (property of how quickly the water saturates the foam) of 6.74 mm/s while the harder foam had 0.84 and 5.66 mm/s respectively. Compression tests were done using an Instron 1000 materials testing machine. The results confirm the idea that there is an increase in load deformation from the addition of water. Both the soft and hard foams exhibited greater strain characteristics under the same stress. The soft foam seemed to "saturate" more quickly from added water than did the harder foam. There were two main implications from this testing. First, if a portion of a cushion became wet, then the load taken by that part was greatly reduced, and the other areas had an increase in pressure. Second, if the whole cushion became equally damp, there was a good chance of "bottoming out".

From these results, it is clear that cushion covers are quite necessary for foam type materials. The cover should have the characteristics of being waterproof, breathable, protective from outdoor elements, and very compliant. A common, commercially available cover in use is made from knitted polyester. It has two-way stretch and adapts to the patient's contours. It also allows water vapor and air to escape from the cushion. If laminated on the inside, it can also act as protection against spills and incontinence.

5.3.2 The hammock effect

The necessity of foam cushion covers leads to concern about the so-called hammock effect. This effect is essentially an expression of the restoring forces generated by tension along a membrane which is loaded by some mass. Figure 5.14 demonstrates how this principle is in effect when sitting on a covered seat cushion. First, figure 5.14(a) shows the typical deformation of a foam type cushion with a cover. Assuming the weight of the patient can be represented at the center of the cushion, there are necessarily reaction forces F which must traverse the cushion cover membrane as it would in the case of a hammock. These forces can be broken down into vertical and horizontal components, F_x and F_y at any point along the cover. If a cushion cover is not very extensible, then the foam underneath will not act as the only primary support (assuming $F = kx$ linear spring characteristics). There will be additional forces F_y which will increase pressures at the seat interface. This is also a problem for gel or fluid-filled cushions that require sealed covers.

Figure 5.14 The physics behind the hammock effect. (a) The deformation of a patient sitting on a covered seat cushion. (b) The forces applied to the closed system of the cushion cover.

Denne (1981) suggested that the hammock effect does not have as much influence as initially expected. He proposed that tension from a cushion cover arises in a different fashion than from a hammock because the cushion cover is not tethered at two points as in a hammock. Using various covers of different rigidity, Denne found that there was virtual absence of any change in the IT pressures with the increase in cover stiffness. He repeated pressure measurements to provide a measure of reproducibility. He found the largest increase in pressure in the two way stretch fabric which exhibited pressures 13% above the baseline pressure. The cushion used for the testing was a standard Department of Health and Social Security (DHSS) polyurethane foam insert. The study did not, however, investigate the presence of horizontal shear forces which may be just as undesirable. It also did not address the issue of whether the membrane of a fluid-filled cushion can have such a "hammock effect." Figure 5.14(b) shows there can be considerable force translated along the cover of the cushion. Denne did note that people who sat on the different covers observed a difference in sensation for the stiffer cushion covers. This might indicate that the IT pressure is not the only factor to consider in making a quantitative analysis of the hammock effect.

Chung (1987) reported that shear and surface tensions (the hammock effect) accounted for 40% and 70% of the total load through load-deflection testing on different foams with a 20 cm and 7 cm diameter indentor respectively.

5.3.3 Thermal factors

While cushion covers protect the cushion from moisture, stain, and soil, they also affect the thermal properties of the cushion. Likewise, different materials exhibit different values of heat flux. Chapter 1 indicated that an increase in temperature can increase the metabolic demands of the blood cells and affect oxygen consumption. On the other hand, cushions that act as a heat sink will be very chilly to sit on and might initiate a vasoconstrictor response evidenced by a further decrease in blood supply (Mooney *et al* 1971).

Various studies have come to a similar conclusion about the thermal properties of the basic cushion materials: foam, gel, air, and water filled. Seymour and Lacefield (1985) found that air-filled and foam cushions would increase skin interface temperatures while gel and water cushions would lower them. Fisher and Kosiak (1979) specifically cited the Roho cushion as increasing the IT temperature 2.3°C and 1.8°C (with and without cover) for a 30-min sitting period. Figure 5.15 shows the change in temperature with

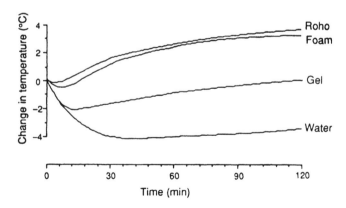

Figure 5.15 Temperature changes in skin from original temperatures for subject sitting on various seat cushions up to 2 h (Lee 1985).

duration of sitting. Lee (1985) found similar evidence in support of these conclusions. The results indicated that foam cushions are "hot" because they are relatively poor absorbers and conductors of heat. Gel pads eventually maintain skin temperature as in air. Water flotation pads provide the most significant amount of temperature drop that is permanently maintained for long sitting durations.

5.3.4 Humidity and ventilation

In addition to temperature, the humidity at the interface can also cause skin maceration and discomfort. However, humidity is quite difficult to measure even with electrical devices. Cochran tried to conduct experiments using sensors made by Humidial Corp. (Colton, CA) which ultimately indicated humidity by color changes on a piece of blotting paper. The sensor can measure relative humidity from 20 to 90%. The results showed that humidity was lowest (15%) for the foam cushions and highest for the gel, water, and air cushions (36.5% average). The evidence indicates that all cushions employing water-impermeable surfaces create greater humidity because moisture is trapped at the interface, whereas porous substances such as foam allow the skin to "breathe." Foam does not encourage humidity increase but does allow higher skin temperatures at the interface.

5.3.5 Pressure sensitivity for air-filled cushions

When using air-filled cushions (AFCs), patients must understand the sensitivity of the cushion to the amount of air pumped into it. There is an optimal range of inflated pressure that will provide the least amount of interface pressure for a particular patient (Krouskop *et al* 1986b). Figure 5.16 shows how under- or

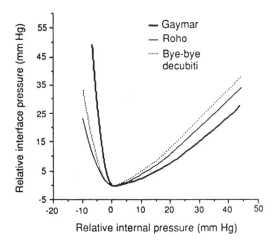

Figure 5.16 Relative interface vs. relative internal air pressure for AFCs. The zero pressures are with reference to minimal achieved interface pressure for differently weighted patients and the corresponding internal pressure at that point (Krouskop *et al* 1986b).

overinflation of AFCs can cause greater than optimal pressures at the interface. Underinflation, in particular, causes a more dramatic increase in interface pressure due to bottoming out. Overinflated cushions create a harder surface because the air has nowhere to be displaced, but the change is more gradual with overinflation. This type of sensitivity would suggest that all AFCs should be required to have a pressure indicator attached to monitor adequate inflation. If this is not possible, Williams *et al* (1983) suggested using a universal inflation pressure of 35 mm Hg for all individuals. Their conclusion came from the results of a test based on what percentage of 14 patients had interface pressures 8 mm Hg within their optimal (lowest) pressure for any given inflation pressure. At 35 mm Hg, 85.7% of the individuals had interface pressures within 8 mm Hg of their lowest possible interface pressures. Although these results offer a general standard for cushion inflation, AFCs should be recommended to patients who demonstrate a high level of responsibility so that proper interface pressures will be maintained.

5.4 OBJECTIVE TEST METHODS FOR EVALUATION

In testing seat cushions, there are several clinical methods for comparing their efficacy. Each method highlights a particular characteristic or function of the cushion. Quantifiable tests are not the basis for ranking seat cushions overall, but the results of each parameter can suggest a cushion's general performance.

5.4.1 Pressure lowered at ischial tuberosity

A large percentage of pressure sores occur near the ischial tuberosities. It is common to principally evaluate cushions on how well they reduce pressures at that area. Several studies provide data on the mean interface pressures of different cushions. The pressures range from 58 to 93 mm Hg, which is still well above the recommended pressure of 32 mm Hg (Houle 1969). Although the three studies shown in figure 5.17 are not altogether consistent, there are some general similarities. All studies indicated that the gel type cushion has the highest pressures. This could be associated with the hammock effect of the membrane holding the gel. Fluid flotation and foam cushions appear to produce consistently lower pressures. Air type cushions yield average-to-low pressures

Cushion type	IT pressures from Cochran and Palmieri (1981) study	IT pressures from Souther study (1974)	IT pressures from Mooney *et al* (1971) study
Foam	67 mm Hg	58 mm Hg	68 mm Hg
Gel	93 mm Hg	67 mm Hg	80 mm Hg
Fluid flotation	71 mm Hg	41 mm Hg	70 mm Hg
Air	79 mm Hg	53 mm Hg	73 mm Hg

Figure 5.17 Mean pressures under the ischial tuberosities for the four basic cushion types.

in each of the studies. Sprigle and Chung (1989) indicated that custom contouring a foam cushion lowers pressures even more: for a high-resiliency polyurethane cushion, contouring improved pressures from 53 to 45 mm Hg.

5.4.2 Pressure redistribution

Pressure gradients across the buttocks can create detrimental shear forces. It is important to observe globally how a cushion distributes pressure over the entire cushion interface. Custom contoured foam cushions and air cushions are noted for their ability to evenly redistribute pressures away from the ischial tuberosities. The VASIO cushion design deliberately places greater pressure on the outer thigh region (figure 5.6) since there is more tissue to support excessive pressures. There has been little research done in the area of pressure distributions because of the difficulties in obtaining accurate data. A useful device in the future would be one that measures pressure distribution similar to an ischiobarograph while being applied at the patient–cushion interface.

5.5 SUBJECTIVE METHODS FOR EVALUATION

The subjective evaluations of patients who use seat cushions are also considered a measure of a cushion's success. Patients may choose a purely aesthetic, comfortable cushion over one that has been clinically effective in reducing interface pressures, even though it may not be adequate for preventing pressure sores. A marketable cushion must have a practical design that provides enough freedom for a patient to use it consistently.

In the SCI seating clinic, most of the evaluations include interface pressure measurement, transfer ability, propulsion, spasticity, balance, posture, skin reaction, pressure reliefs, and comfort.

5.5.1 Stability and handling

There are many active spinal cord injured persons who require a stable or firm cushion as well as one that is lightweight. Using a heavy or unstable cushion makes it difficult for patients to transfer in and out of their wheelchair or to a vehicle since the cushion remains with them all the time. Many paraplegic persons do not have good muscular control of their upper body, and it is necessary to have a cushion that will not shift its center of gravity. Liquid-filled and air-filled cushions are difficult to move because they do not have a fixed center of gravity. Gel cushions can be quite heavy and cumbersome.

5.5.2 Cost

The patient is the one who must ultimately buy the cushion, and cost is always an important consideration. Unfortunately, there is no definite correlation between price and efficacy of cushions. Figure 5.18 compares some of the prices of commercially available seat cushions. Foam type cushions have consistently been the lowest in cost because of the relatively simple design. Gel combination cushions such as the Jay incorporate some contouring and special designs to prevent leakage, and are more expensive. Water and air type cushions fall into a medium price range.

Cushion name	Manufacturing company	Material type	Suggested retail price
Jay®	Jay Medical, Ltd.	Gel and Foam	$295–$400
Roho®	Roho Inc.	Air	$300–$360
Hydro-Float®	Jobst	Water	$165–$197
Temper Foam®	Kees Goebel Medical	Foam	$27–$108

Figure 5.18 1990 costs for common seat cushions representing the four major material types.

5.5.3 Maintenance and repair

A seat cushion should be as durable as possible and ideally maintenance free. Unfortunately, there is currently no cushion that satisfies both criteria. Foam cushions are more maintenance free than any fluid-filled cushion and they are not rendered useless from a small tear or puncture. But most foams have a short lifetime of about six months before their mechanical properties begin to change. Air-filled cushions have the special problem of maintaining proper inflation pressures.

5.6 STUDY QUESTIONS

5.1 Sketch the cross section of a Roho cushion. Explain its advantages and disadvantages.
5.2 Sketch a force diagram and explain the source and results of the "hammock effect."
5.3 Explain the advantages of a gel cushion over a water, air, or foam cushion.
5.4 Explain how particle-filled cushions are used in developing support surfaces.
5.5 Describe procedures for fabricating custom contoured cushions.
5.6 Explain the choice of cushion cover for continent and incontinent patients.
5.7 Plot the maximal allowable pressure vs. location in the pelvic region.
5.8 Discuss the need for stability and the stability provided by different support surfaces.
5.9 Explain why underinflation is worse in figure 5.16.
5.10 Describe the philosophy behind the designs of the contoured slab cushion (with cutout and bar) and the VASIO-P cushion.
5.11 Explain how a cushion is developed using computer-facilitated design. Give the advantages and disadvantages of this procedure.

6

Other Support Surfaces

Shen Luo

In order to prevent pressure sores, we wish to obtain an even pressure distribution and decrease time at risk. Chapter 5 provided a detailed discussion of seat cushions. This chapter will continue to study other support surfaces. First, we will discuss wheelchair design and modification, including some special cases. Then we present patient support surfaces for both the hospital bed and the operating table.

6.1 GENERAL CONSIDERATIONS

Paralyzed patients spend their lifetime in a wheelchair or bed. One of the major complications facing them is pressure sores. Clark *et al* (1978) reported that the incidence of pressure sores is 21.6% for paraplegic persons and 23.1% for quadriplegic persons.

A pressure sore is an ulceration of skin and/or deeper tissue due mainly to unrelieved pressure, including shear pressure. Bailey (1967) presented four methods of prevention: (1) produce weightlessness, (2) reduce body mass, (3) increase bearing area, and (4) decrease time at risk. Method (1) is not practical. Method (2) may actually increase the pressure over bony prominences because of reduced padding over them (Barbenel 1987). Also, it is not available for thin and weak patients.

In method (3) we use the "zero pressure" part of the body, where no pressure is normally applied, so that weight is distributed as widely as possible. As an alternative to method (3), we often try to produce even pressure distribution by decreasing pressure at higher pressure areas where patients frequently develop pressure sores and dispersing the forces to other lower pressure areas of the body.

Figure 6.1 shows how a cushion increases the bearing surface and makes the pressure distribution more uniform.

A way of decreasing the time at risk is by relieving the pressure regularly. This is necessary but also difficult to carry out. Paraplegic persons should be trained to perform regular push-ups and tetraplegic patients forward/side leans (Chapter 12).

For patients confined to bed, regular turning is an effective way of reducing the risk. Even for patients in the surgical environment on an operating table, regular relief is the most useful method of pressure sore prevention although this may not be common practice.

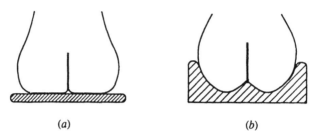

(a) (b)

Figure 6.1 Profile of seating on different support surfaces. (a) A firm surface and (b) using a cushion. A cushion both increases bearing area and improves pressure distribution, especially by reducing the maximal pressure under the ischial tuberosities.

6.2 WHEELCHAIR DESIGN

Wheelchair patients frequently develop pressure sores on the buttocks because of sitting too long without lifting up to relieve the pressure. Mawson *et al* (1988) showed that between 32% and 40% of Spinal Cord Injured (SCI) persons developed a pressure sore within their lifetime. El-Toraei and Chung (1977) and Constantian and Jackson (1980d) found that ischial pressure sores are the most prevalent. Therefore a major topic for wheelchair design is how to design the wheelchair support surfaces to minimize ischial pressure. The Department of Veterans Affairs (1990) has published a 118-page document: "Choosing a wheelchair system," that has 11 index entries for pressure sores.

6.2.1 Differences between able-bodied and SCI persons

Chapter 1 described the physiological changes that take place in SCI persons. Now we describe the differences in pressure distributions between able-bodied and SCI persons.

Most of the research on pressure distribution on seat cushions uses able-bodied persons as subjects. But SCI persons do not have the same pressure distribution.

The spine and pelvis form a vital support system to protect internal organs from being injured, and as a protective measure, the muscles in the area of the bony prominences increase the weight-bearing area. But for SCI persons such muscle has atrophied. Thus the bones protrude more.

In fact, Hobson (1989) has identified pressure distribution differences between able-bodied and SCI persons. The study defined nine typical wheelchair sitting postures and then applied them to two groups of subjects, one of able-bodied persons and the other of SCI persons. Results showed persons with a SCI have mean maximal pressures that are significantly higher than those of able-bodied persons in all nine sitting postures, and the differences range from 6% when the back reclines 30° from vertical posture to 46% in the forward flexed 30° posture. Figure 6.2 compares mean maximal pressure in three sitting postures out of nine, including minimal and a maximal differences as well as a neutral position. We use maximal pressures rather than ischial pressures because change of sitting posture causes a shift of location of maximal pressures. Forward trunk flexion 30° with respect to the neutral posture results in, for instance, a shift of 2.4 cm.

Postures	Mean maximal pressure of SCI compared to able-bodied
Forward trunk flexion 30°	46% higher
Neutral position	26% higher
Back recline 30° from vertical	6% higher

Figure 6.2 A comparison of mean maximal pressure between SCI and able-bodied persons in three sitting postures (Hobson 1989).

Hobson (1989) also observed peak pressure gradients, defined as the maximal gradient between any two adjacent pressure recording cells, of persons with SCI 1.5 to 2.5 times larger than those of able-bodied persons. This implies that there is higher pinch shear stress (Section 2.1) because of higher pressure gradient and therefore may contribute to pressure sore formation for SCI persons.

Moreover, asymmetrical ischial loading in the sitting posture is almost a universal problem for SCI persons due to spinal and pelvic deformities, spinal scoliosis, and pelvic obliquity, caused by the paralysis of musculature and/or the damage of nerve, which results in higher pressure on the lower ischial tuberosity.

6.2.2 Wheelchair adaptation

Correct seating
Method (3) of Section 6.1 suggests that the center of gravity of the torso should be anterior rather than over or posterior to the ischial tuberosities when sitting. The posterior position makes the coccyx become weight bearing. Another reason for anterior posture is based on the study of Reswick and Rogers (1976) which indicates that the outer regions of the buttocks can sustain greater loads than the ischial tuberosities. In addition, the line of gravity should fall in the center of the buttocks (Moe *et al* 1978). Therefore, correct seating is a symmetrical upright sitting posture characterized by a forward-rotated pelvis and a lordosis lumbar spine (Akerblom 1948, Zacharkow 1984, Settle 1987) as shown in figure 6.3.

Also, advantages for such a sitting posture include reduced lumbar disk pressure (Andersson *et al* 1975), relaxation of the lumbar paraspinal muscles (Akerblom 1948), improved diaphragmatic breathing (Bunch and Keagy 1976), and less fatigability (Young and Burns 1981, Grandjean *et al* 1969).

Figure 6.3 A correct seating position is an upright sitting posture with lumbar lordosis. Adapted from Zacharkow (1984).

Sling seat

Considering the requirement of portability, a majority of wheelchairs are designed to be folded. A sling seat and backrest are standard on wheelchairs and cause inherent problems.

Houle (1969) found that the sling seat eventually leads to very asymmetrical sitting with able-bodied persons. For SCI persons, the sling seat invites (Gibson and Wilkins 1975) and/or aggravates spinal/pelvic deformities (figure 6.4) as mentioned in Section 6.2.3. This implies that correcting spinal scoliosis and pelvic obliquity on a sling seat is almost impossible.

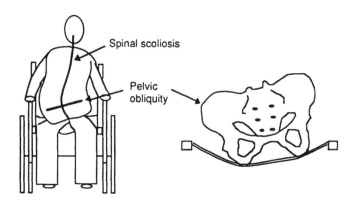

Figure 6.4 The sling seat may cause spinal scoliosis and pelvic obliquity.

One modification is to use a plywood board over the sling seat. To retain portability, the board is fit between the side rails and attached by Velcro to the sling as shown in figure 6.5 (Zacharkow 1984). Another modification is to employ a very firm foam base and a modular design (Chapter 5).

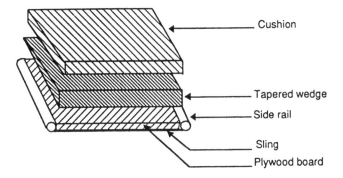

Figure 6.5 Adding a plywood board over a wheelchair sling minimizes spinal deformation.

Sling backrest
With a sling backrest, a typical wheelchair causes a kyphosis of the lumbar spine, posterior sitting posture with the center of gravity of the trunk over or behind the ischial tuberosities (figure 6.6).

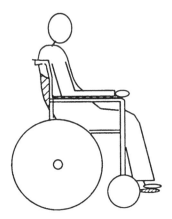

Figure 6.6 A sling backrest in a wheelchair causes kyphotic posterior seating posture.

Koreska *et al* (1976) and Wilkins and Gibson (1976) note that kyphotic seating leads to pelvic obliquity. Therefore, the sling backrest must be as tight as possible and be reinforced to eliminate any sagging.

Wheelchair angles

SCI persons have two problems when they sit in a horizontal seat. As in figure 6.6 with a sling backrest, there is the tendency for the torso to slide downward and the buttock and thigh forward, which results in not only the high pressure behind the ischial tuberosities but also shear pressures and friction. Second, this is a unstable position because most SCI persons lose their balance and fall forwards when sitting upright.

Zacharkow (1984) recommended a wheelchair with 10° seat and 15° backrest inclination shown in figure 6.7, and suggested these are good compromise angles for both functional activities and relaxation.

Gilsdorf *et al* (1990) used a forceplate mounted on a wheelchair seat to measure normal and shear force. Results suggest that the wheelchair user should momentarily lean forward after a recline to reduce undesired forces.

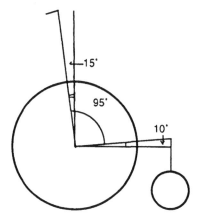

Figure 6.7 Wheelchair angles recommended by Zacharkow (1984). For a C4-spared quadriplegic person, both seat and backrest inclination may need to increase 5° for better balance.

The modifications in most wheelchairs are employing tapered wedges with very firm foam for the seat (figure 6.5) and/or backrest inclination.

Lumbar support

To achieve correct seating, Zacharkow (1984) designed a lumbar support or pad with very firm polyurethane foam and placed the lowest edge of the lumbar pad slightly below the highest part of the posterior iliac crests as shown in figure 6.8. Compared with figure 6.3, this sitting posture with wheelchair adaptation matches an upright sitting posture with lumbar lordosis as closely as possible.

Shield (1986, 1987) studied this adapted wheelchair on twenty able-bodied persons with and without lumbar support. The studies were taken on a hard surface so an ischiobarograph (Chapter 10) could measure the pressure distribution. The result showed a large percent decrease of high pressure areas greater than 500 mm Hg and a significant increase of the area in the pressure interval less than 100 mm Hg with lumbar support.

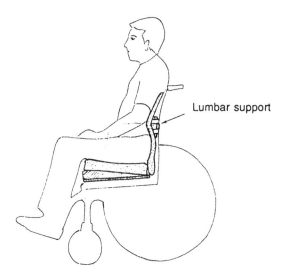

Lumbar support

Figure 6.8 The adapted wheelchair with lumbar support. This seating matches the correct seating shown in figure 6.3 as closely as possible (adapted from Zacharkow 1984).

For preventing pressure sores, of course, many details of wheelchair design must be considered. A proper wheelchair for an individual should have proper armrest height, sling width and depth, backrest height, and footrest height. Settle (1987) suggested that the team for wheelchair adaptation should fully cooperate with the local Artificial limb and Appliance Center and with a competent orthopedist because adapting wheelchairs for patients with seating problems is often a difficult and time-consuming task.

Armrest height
An armrest can support the upper part of the torso. According to Brattgard and Severinsson (1978), a proper fitting armrest can reduce pressures under the ischial tuberosities by about 25–30%.
    If too low or removed, there will be no support function. Therefore, adjustable and well-padded armrests should be a standard feature on all wheelchairs (Zacharkow 1984).

Sling width
Too wide a chair for an SCI person, as mentioned before, will invite spinal scoliosis and pelvic obliquity; too narrow a chair requires regular push-ups to relieve pressure. A proper width wheelchair should be ordered.

Sling depth
For tall patients, a relatively short seat sling causes a reduction of the pressure-bearing area of the thigh. For short patients, a relatively long sling results in the posterior seating posture with lumbar spine kyphosis, and high pressure over the coccyx and sacrum, along with a shear force over the same areas.

Backrest height
Because a wheelchair-dependent person spends 16–18 h/day in a wheelchair, a wheelchair should be designed for both work and rest. A relaxing backrest should support the thoracic region of the back (Diffrient *et al* 1974). However, a working backrest should not cover the scapulae (Floyd and Roberts 1958), because it interferes with arm motion such as propelling the wheelchair. The best compromise is that the backrest should extend to approximately 1.3–2.5 cm below the inferior angle of the scapulae (Zacharkow 1984)

But for an SCI person with poor balance, a constant effort to pull the head and shoulders forward will cause severe increased spasm so that he gives up trying to relieve his own pressure due to exhaustion. Therefore Settle (1987) suggested that a detachable backrest extension be provided for those persons.

Footrest height
Too low a footrest will cause a pinch shear (Chapter 2) and add the weight of the lower legs to the thighs. Too high a footrest will add almost one-half the weight of the thighs to the buttocks, which results a large increase of ischial pressure. Therefore, a proper height for the footrest is very important. Gilsdorf *et al* (1990) used a forceplate mounted on a wheelchair seat to measure normal and shear force. Results suggest that wheelchair cushions with firm material under the thighs will facilitate reduction in ischial tuberosity pressure when the leg height is lowered as much as possible. Angles of the footrest are also important to stabilize the lower extremities, particularly for SCI persons and others with sensory and motor loss.

6.2.3 Special design

As mentioned in Chapter 5, the particle filled cushion (PFC) has been recommended for particular patient groups: spastic paraplegia and tetraplegia, cerebral palsies, and patients with asymmetric shape due to spinal and hip deformities. The VASIO cushion design (Chapter 5) may be another way. Here we further introduce other measures for these groups of patients.

Patients with C1 to C6 spared
Due to the paralysis of the trunk musculature and the effect of gravity, individuals with a complete cervical or high thoracic spinal cord injury (C1 to C6 spared) show a thoracolumbar C-curved scoliosis with pelvic obliquity when sitting, with the lower side of the pelvis on the convex side of the curve (Zacharkow 1984).

McKenzie and Rogers (1973) recommended a trunk support to stabilize the thoracic spine (figure 6.9). For supporting the lumbar spine, an abdominal binder is needed.

Patients with muscular imbalance
Trunk muscular imbalance, which is frequently due to a symmetrical damage or regeneration of the lumbar nerve roots, can result in a lumbar or thoracolumbar scoliosis with pelvic obliquity. Then the person cannot level the pelvis in a supine position (Mayer 1936).

Figure 6.9 A trunk support stabilizes the thoracic spine for C1 to C6 spared.

The goal toward correcting it is to make the line of gravity fall in the center of the buttocks. Therefore Zacharkow (1984) recommended a gluteal pad and side supports as shown in figure 6.10.

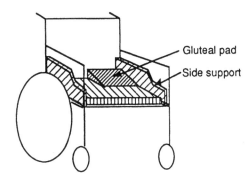

Gluteal pad

Side support

Figure 6.10 Gluteal pad and side supports correct trunk muscular imbalance.

Patients with cerebral palsy
For patients with cerebral palsy, wheelchair design often is more complicated. A seating system must accommodate a wide range of muscular and/or orthopedic related problems.

Ginpil *et al* (1984) design seat, foot, and backrest supports using an adjustable mesh (figure 6.11). The fitting procedure is (1) to permanently sew all straps at one side of the frame and to temporarily clamp them on the opposite side; (2) to change the lengths of the individual straps when the person is sitting; (3) to remove the person and permanently sew the free end.

Cooper and Hawkes (1984) developed a shapeable matrix support surface. The basic components include a cylindrical node with 8 slots and 7 types of beams (figure 6.12).

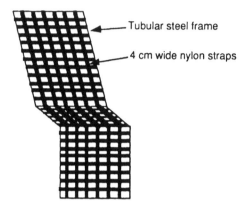

Figure 6.11 The adjustable mesh support surface has individually adjustable straps.

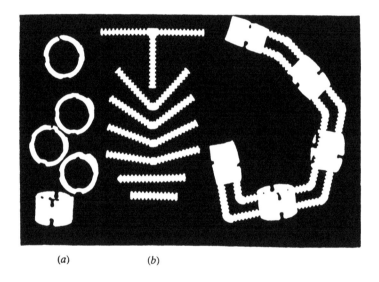

      (a)   (b)

Figure 6.12 The shapeable matrix support surface contains (a) nodes and (b) beams which snap into the nodes. Two beams are used between each node. A typical seat matrix contains 150 to 250 nodes and 400 to 600 beams.

The assembly procedure involves initial measurements, matrix construction, and subject fitting with repeat measurements. A skilled technician needs 4–8 h to complete a seat. The matrix support surfaces are shaped to help maintain pelvic stability and avoid pressure in the areas of the great trochanter, sacrum, and gluteal cleft.

6.3 HOSPITAL BED

Unlike wheelchair-dependent patients, major risk areas for bedridden patients are the sacrum, the heels, the scapulae and the occiput when lying in the supine position and the trochanter when in the lateral position (figure 6.13).

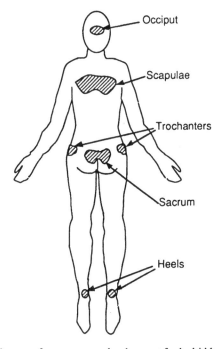

Occiput

Scapulae

Trochanters

Sacrum

Heels

Figure 6.13 High risk areas of pressure sore development for bedridden patients.

Bailey (1967) reported that a 48-year-old man developed bilateral ischial sores in his wheelchair. During several weeks of bed rest, as the ischial sores healed, the sacral and trochanteric regions broke down. This suggests that decreasing the pressure of these bony prominences is the major task of bed design for bedridden patients.

6.3.1 Ideal hospital bed

A ideal hospital bed should:

(1) provide even pressure distribution
(2) provide no shear forces
(3) provide proper temperature and humidity for skin
(4) be comfortable
(5) be economical
(6) be accessible for care

A practical bed only satisfies some of the requirements (Torrance 1983).

6.3.2 Mattresses

When lying on a firm support surface, the majority of the body's weight is borne by five bony prominences, that is the sacrum, heels, scapulae, occiput, and trochanters.

Like seat cushions for wheelchairs, mattresses are employed in order to increase the weight-bearing area and produce as even a pressure distribution as possible.

Comparative studies for many different kinds of mattresses indicate that there is great variability in mattress effectiveness.

Garfin et al (1980) using 65 water-filled bladders (Chapter 8) studied the surface pressure distribution of four mattresses: (1) an orthopedic 720-coil mattress with built-in bed board and 13 mm of foam on top (Simmons Hardline, Simmons Co., Los Angeles, CA); (2) a standard 500-coil bed (Royalty 510, Levitz Furniture Corp., Miami, FL); (3) a commercially available water bed filled to a depth of 25 cm (Watercloud Super Supreme 2653, Watercloud Bed CO., Santa Ana, CA); and (4) a hybrid bed combined a shell of polyurethane surrounding 8–10 cm of water (Aquapedic GDS Water Mattress, Aquatherm Co., Rahway, NJ). The results in the supine position show the hard 720-coil bed and the commercial waterbed distributed weight more evenly than the other structures (figure 6.14), involving almost the entire body surface in the weight-bearing pattern.

Krouskop et al (1985a) evaluated the pressure-reduction characteristics of several mattress overlays. A hospital bed with a standard mattress was used as a control surface. The results shows the Roho and Akros mattresses are more effective overlays, whereas a 50 mm thick convoluted foam provides no significant protection for the trochanter when lying in the lateral position (figure 6.15).

Wharton et al (1988a) compared two kinds of foam mattresses with three other kinds of mattresses using the TIPE pressure measurement system. He found that the Kin Air Mattress had significantly lower average pressure relief characteristics than any of the other mattress types, and that the foam products were similar to the other two kinds of mattresses: Geo-Matt and Gaymar Soft-Care Mattress.

The above three studies were all based on able-bodied persons. Therefore, as in the investigation for wheelchair-dependent persons (Section 6.2.1), it is necessary to study whether differences occur between able-bodied and SCI persons.

In general, it is considered that pressures greater than 30 mm Hg can cause tissue ischemia. Therefore, bedridden patients should choose a mattress with the most even pressure distribution to help minimize the chance of pressure sore development. Most commercially available mattresses yield in excess of 30 mm Hg. Also, it is always necessary that mattresses permit regular turning (Chapter 7).

6.3.3 Distribution and weight

Barbenel (1987) studied the pressure distribution of persons on mattresses and found that persons can be classified by their weight. The lightest subjects produce the highest pressure.

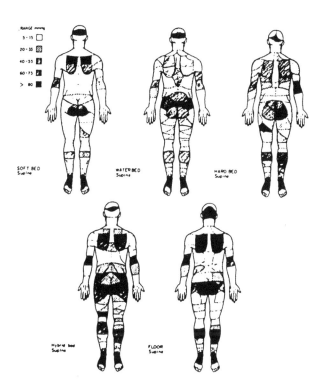

Figure 6.14 The mean pressure distribution of the hard bed and the water bed were most uniform (Garfin *et al* 1980).

Mattresses	Scapulae	Sacrum	Trochanter
Standard	31	33	73
Roho	23	22	45
50 mm convoluted foam	29	27	66
100 mm convoluted foam	23	25	51
Akros DFD support	20	20	45

Figure 6.15 The Roho and Akros mattress overlays provide lowest maximal pressures in mm Hg (Krouskop *et al* 1985a).

Solis *et al* (1988) reported that most pressure sores in recumbent children occur in the occipital region, compared with only 1% of all pressure sores in the adult population. This may be because a child's head has a greater percent of body weight than an adult's.

6.3.4 Distribution and comfort

Garfin *et al* (1980) thought that an even distribution of body weight with the bed contributing support over a large surface area might give a person a more restful period while recumbent.

But Krouskop *et al* (1985b) found that mattresses with uniform pressure distribution make the person restless. He found that the interface pressures that are required to maintain comfort are incompatible with the interface pressures necessary to minimize the risk of creating soft tissue breakdown. As a compromise, he thought that the most effective method was by using support surfaces that had varying stiffness to provide the shear that was necessary to let the body's musculature relax while maintaining skeletal alignment.

But for sitting postures, Shield (1987) thought that because of the serious consequences resulting from a pressure sore, a posture that potentially enhances the development of a pressure sore could never be justified regardless of other conceivable benefits. A notable fact is that astronauts in space can tolerate no pressure distribution for months or even a year. It may be necessary to train persons to adapt to the mattresses that yield even pressure distribution.

6.3.5 Minimizing shear

Chapters 2 and 11 present the causes and mechanics of shear. Here we present two ways to minimize shear force for an elevated hospital bed. A first is use of a bed with elevation of the thigh section in order to prevent the torso from sliding down. A second is to use the Roho balloon pads (Chapter 5) or mattress. As when applied for direct pressure, the Roho overlay minimizes maximal shear (e.g. in the sacro-coccygeal region when lying on a elevated hospital bed) by means of producing more uniform shear distribution. But it may cause sliding due to simultaneously decreased pressure at the same area (Chapter 11). A combination of the above two measures may be the best solution for the elevated hospital bed.

6.4 OPERATING TABLE

Patients during a surgery of many hours may acquire pressure sores that appear postoperatively.

Many factors contribute to pressure sore formation. Some chronic or critical patients have been at high risk and surgical interventions of more than 3 h induce and/or exacerbate the development of pressure sores. Snell (1989) reports that 15 out of 72 patients undergoing elective total colectomy with continent ileal reservoir reconstruction in the dorsolithotomy position for 8–13 h developed Stage II–III pressure sores 48–72 h after the operation, and 12 of the 15 became chronic.

The normal protective processes of discrete movement associated with ischemic discomfort present in normal waking and sleep patterns are obliterated during deep levels of surgical anesthesia. Sudden perfusion, such as rapid intravascular fluid administration or sudden release of the source of external pressure, can cause local capillary collapse and subsequent ischemic insult and, oftentimes, necrosis.

In order to accommodate the optimal surgical access and to obtain proper physical support, the operating table is adjusted to various surgical positions. The specific weight-bearing regions are at risk of pressure sore formation. In fact, the standard operating table that is constructed with a 25–50 mm foam mattress offers little protection to anesthetized patients.

Relieving excessive pressure regularly may be necessary. Currently available alternating air pressure mattresses designed for standard hospital beds (Chapter 7) are not available for use on the operating table. They are so wide they would interfere with the surgical team as they lean against the table. Noise is another problem. Infrequently, the mattress is inadvertently punctured and deflated.

Snell (1989) has developed a doubly staggered computerized sequential inflation device. A improved alternating air pressure mattress (a Kendall product) is placed between a commercialized heating blanket and the waterless gel pad (Chapter 5) from the level of the ischial crest to the sacrum. Initial results show that it can reduce the severity and number of Stage II–III ischemic lesions.

6.5 BED TRACTION

When bones are fractured, the orthopedic department may apply bed traction (orthopedic traction, traction table). Traction overcomes (1) the deforming effect of force, (2) the effects of muscle tone, and (3) the effect of the skeleton so that normal alignment is restored (Lewis 1977). There are two basic types of traction, manual and continuous. Here we only discuss continuous traction, which is divided into skin traction (figure 6.16(a)) and skeletal traction (figure 6.16(b)).

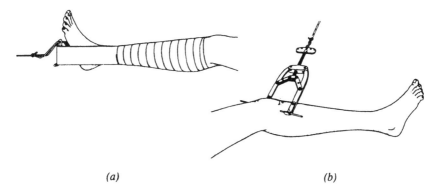

(a)   (b)

Figure 6.16 (a) Skin traction and (b) skeletal traction.

In continuous traction an applied force must remain constant in magnitude and direction until the fracture fragments unite. This period can be weeks or even months. The possibility of developing pressure sores is a inherent problem. Petersen (1976) reported 11% of pressure sore patients developed their sores during treatment of fractures of the femoral neck because the traction apparatus applies pressure to a small area extending from the sacrum to the perineum. He indicated that bed traction is more dangerous to the patient than the operating table.

In fact, prolonged direct pressure and asymmetric shear force may be the main contributor to pressure sore formation. Stone and Lambert (1975) indicated it is very important that the patient be able to move about without changing the position of the part to which traction is applied. To supply this freedom of movement, a second system of weights and pulleys called "balanced suspension" is often used. Balanced suspension permits the limb to "float" over the bed and also facilitates bedpan use and changing of bed linen with minimal disturbance of the fracture.

Meanwhile, Stone and Lambert (1975) indicated that the patients should be placed on a firm mattress which does not sag under the weight of their bodies. From study of Garfin et al (1980, figure 6.14), an orthopedic 720-coil mattress with rather even pressure distribution is just such a hard bed. On the other hand, a Low Air Loss Bed System (Watkins and Watson Ltd, figure 7.3) is also equipped as a traction bed.

Lewis (1977) thought when patients complain of a burning pain in the area that has undue pressure, these complaints should be investigated thoroughly to prevent pressure sores. Moreover, good supervision is required. Some companies (e.g. Zimmer·USA, K&M Company, etc) also emphasize in their manuals avoiding pressure at some particular locations when setting up traction apparatus, especially at bony prominences.

There are large shear forces on the skin surface when applying skin traction. Therefore the pulling force should not be applied with too large a magnitude and duration.

6.6 SPINAL BOARD

In the United States alone, 7000 to 10000 people have a traumatic spinal cord injury every year. A majority of spinal cord injuries are caused by motor vehicle accidents (47.7%). Falls (20.8%), acts of violence (14.6%), sports (14.2%) and other activities (2.7%) account for the remainder of spinal cord injuries. Diving and football are the major sporting activities in which people have been spinal cord injured. Most spinal cord injured persons are male (82%) and between 16 and 30 years of age (61.1%)(Buchanan 1987).

Spinal boards are used for two reasons: (1) many times the spinal cord is not completely severed, then some return of function below the injury is possible; (2) a sharp fracture fragment could stab or puncture nearby nerves and blood vessels. To prevent any further damage during extrication and transport, after the likelihood of a spinal cord injury is determined, the injured person will be immobilized and strapped to a spinal board often with the head taped to it at the emergency scene. After arrival at the emergency room, X rays of the person are taken while still immobilized on board.

A spinal board is a bare, hard plywood board about 190 by 40 cm with holes on the four sides to strap patients. There is no soft cushion overlay on it because any jarring to the injured person could cause further insult and sometimes be dangerous (e.g. the edges of fracture could sever blood vessels). Before getting a definite diagnosis, immobilization must be maintained. Depending on transport, examination, diagnosis, etc, the duration can be from one-half hour to several hours, and occasionally 8 hours, according to emergency room staff of University of Wisconsin Hospital and Clinic. Because the main focus is on emergency treatment and the relatively long time that may take place before an actual pressure sore appears, preventing pressure sores has not been a priority. But the possibility of pressure sores exists, and the longer the patients stay on the spinal board, the higher the risk. Therefore, personnel should be concerned with reducing the time of transport and treatment.

Another possibility for simultaneously providing stability and preventing pressure sores might be to design a lightly padded, contoured spinal board to relieve pressure at high pressure points.

6.7 STUDY QUESTIONS

6.1 Define a ideal patient support surface system.
6.2 Describe four direct methods of preventing pressure sores.
6.3 Explain why maximal pressures on spinal cord injured persons are higher than those on able-bodied persons.
6.4 Define correct sitting posture.
6.5 List problems in construction of a majority of wheelchairs.
6.6 Explain how wheelchair design for posture control can reduce interface pressure.
6.7 Describe why and how to adapt wheelchairs for different persons.
6.8 Explain how to implement correct support surfaces for a SCI person when sitting.
6.9 Sketch different methods of forming adjustable support surfaces for wheelchairs.
6.10 Suggest improvements to wheelchair design.
6.11 Explain the location of the most prevalent pressure sores when sitting for SCI persons and how to reduce the risk.
6.12 Explain the location of the major areas of pressure sore development for bedridden patients.
6.13 Explain why the lightest patients produce the highest pressures on mattresses.
6.14 Explain how to minimize shear in a hospital bed.
6.15 Describe the construction of hospital bed mattresses.
6.16 Explain why stiff mattress covers produce higher interface pressures while stiff wheelchair covers do not.
6.17 Suggest improvements to the operating table mattress.
6.18 Suggest improvements to the spinal board.

7

Complex Pressure Relief Methods

Shu Chen

In this chapter we present different kinds of patient support systems such as the Alternating Pressure Mattress (APM), the Low Air Loss Bed System (LALBS), the Air Fluidized Bed (AFB), the Water Bed, the Automated Bed, and the Oscillating Wheelchair. We discuss their working principles, advantages, disadvantages, and applications. We also present experimental data that show how well those pressure relief methods succeed in redistributing pressure on the body.

7.1 ALTERNATING PRESSURE DEVICES

Alternating pressure devices consist of air cells connected to an electrical air pump which alternates the area of the body under pressure. The alternating pressure mattress (APM) is the major example in this group. They can be cycled by either air or water. Their sizes are variable and could be designed either for bed or wheelchair cushions.

7.1.1 Alternating pressure mattress

The APM has been available for more than 30 years. Figure 7.1 shows an overview of the system. The system consists of a cellular air mattress connected to an electric air pump. The mattress is made up of two sets of tubular air cells which are alternately inflated and deflated. The cycle varies from 2 to 6 min depending on the model and is controlled by a time switch in the pump unit. In the operation, the first set of cells inflates, the second set remains deflated; then the second set inflates, until both sets of cells are full; finally the first set deflates. This cycle is automatically repeated so that no part of the body is under constant pressure. Most APMs have a protective cover. The air tubing is connected to the pump unit's outlet and the unit can be suspended at the foot of the bed where its warning lights can be easily seen. Some pump units can operate two APMs and others can be set for large-celled or small-celled mattresses. Most pump units can also be adjusted to an appropriate weight setting. If necessary the patient can be placed on it before it is fully inflated.

However, it is better to allow the mattress to inflate fully before placing the patient on it, as this allows for fault checking and repair. While the mattress inflates two indicator lights will come on. If the red warning light does not go off within 10 to 45 min (depending on the model) it indicates a low pressure and the mattress must be checked.

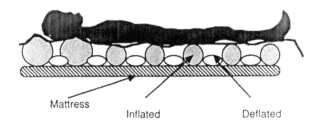

Mattress

Inflated   Deflated

Figure 7.1 The alternating pressure mattress alternately inflates every other air cell so no tissue receives a constant pressure.

There are several different types of APMs. Conventional APMs range from small-celled models (height 3–5 cm) to medium (7 cm) and large (10–11 cm). "Bubble" pads which feature many small, hemispherical cells (6 cm) and dual mattress systems have been developed too. Also, the air cell could run lengthwise instead of crosswise. This gives a smoother appearance and overcomes the tendency of the pad to creep toward the foot of the bed when the back rest is elevated. The large cell mattress is more effective than the medium or small cell types. APMs are light-weight and fit on top of an ordinary mattress. They can easily be used in the home. When used correctly a large-cell APM is an effective aid to pressure sore prevention. It is economical and as the quality of the materials improves should become more reliable. Modern mattresses feature interchangeable cells, so maintenance is much easier and quicker. Many of the problems leading to the failure of an APM result from mistakes in setting up the equipment, poor maintenance, or an insufficient understanding of their working principles. The user must ensure that the pump setting suits the type of mattress and the weight of the patient.

Gardner and Anderson (1954) investigated the use of an alternating pressure pad in prevention and treatment of pressure sores. For five years the pad had been used routinely for patients admitted with pressure sores or who are considered candidates for development of this condition. In analysis of 38 patients with pressure sores on admission 15 showed complete healing, and 19 showed definite improvement.

7.1.2 Fluid pressurized cushion

Schulman (1989) invented a hollow, air-filled cushion (figure 7.2) which is formed from typically three interfacing matrices. Each matrix is a set of hollow bellows-like cells formed from natural or synthetic rubber or rubber-like plastic. The cells of each matrix are spaced apart to accommodate between them cells of each of the other matrices to define a body support surface made up of the tops of all the cells. Each matrix has separate fluid ducts between its cells. Air pumps are used to inflate and deflate the matrices in sequence to shift body support from one set of cells to another for promoting blood circulation and enhancing comfort.

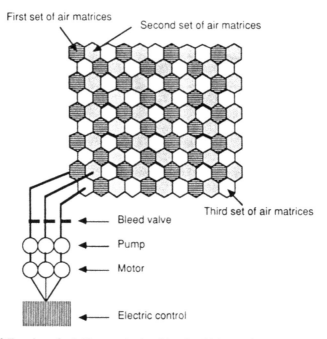

Figure 7.2 Top view of a fluid pressurized cushion, in which one of three matrices is deflated in sequence (Schulman 1989).

Previous inflated cushioning devices have had some shortcomings. In some devices a leak could cause the cushion to collapse, rendering it ineffective. Some cushioning devices were not thick enough to fully contact and support the user's body contours without bottoming out. When cushion inflation pressure was increased to prevent bottoming out, the ability of the cushion to conform closely to the user's skin was reduced. As the cushion became more firm, its benefit to the user decreased. When the cushion was made thicker to improve conformability, it tended to become more difficult for ventilating air to reach the skin and keep it cool and dry and thus increased the risk from maceration. This device can also be used as a highly beneficial passive cushion. Moreover, it will not bottom out even if punctured. Although this device seems promising in preventing pressure sores, we have found no report of clinical use.

7.2 AIR BEDS

Air support systems use the air flotation principle to support patients. The major air support systems we discuss here are the Low Air Loss Bed System, the Simpson–Edinburgh Low Pressure Bed and the Air Fluidized Bed.

7.2.1 The Low Air Loss Bed System

Figure 7.3 shows the Low Air Loss Bed System (LALBS) (Watkins and Watson Ltd, Wareham, Dorset, UK). This is a system where the patient is supported on 21 vapor permeable air sacs of polyurethane-coated nylon, each with a depth of 30 cm. The 21 air sacs are in five groups, each group having its own valve to control the air pressure. The pressure in the groups of sacs can be adjusted when the position of the patient is changed to provide maximal contact with the surface of the body. Variable positions of comfort are obtained by means of simple controls, which enable the contour or attitude of the patient to be adjusted, Temperature-controlled air is supplied by a blower unit. The idea of the LALBS came from a High Air Loss Bed System (HALBS) which was developed to treat severe burn injuries. The HALBS supported the patient directly on air (Scales *et al* 1967). The LALBS was developed from the HALBS for treating the later stages of burns and other conditions which necessitated prolonged bed rest. The LALBS is claimed to fulfill the requirements of an ideal patient support system as proposed by Scales (1976).

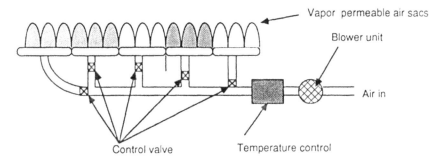

Figure 7.3 The Low Air Loss Bed System feeds temperature-controlled air into a deformable and adjustable water vapor permeable support surface. Unlike the Roho the air sacs run across the full width of this bed (Scales *et al* 1967).

The LALBS requires less than 100 ft^3/min (47 l/s) of temperature-controlled air at a pressure in the range 6 to 33 cm H$_2$O (0.6–3.2 kPa), and permits the transference of water vapor from the skin surface to control the microclimate of the patient. The acceptable temperature range for many patients is 28–31°C. If the temperature of the air is allowed to rise above 35°C many patients start to sweat. The air pressure valves which are sited below the pressure gages on the blower unit control the air pressure in each of the five groups of air sacs. Pressures are correctly adjusted when the air sacs are depressed to the contour of the patient's body. The water vapor from the patient diffuses through the membrane into the air sac, where the relative humidity of the air is on the order of 27%. The shape of the sac permits deformation of the membrane to accommodate the contours of the body without high local pressure areas developing. Two types of air sac membranes are now in use. One is physically porous, the other chemically porous to vapor.

The mobile blower unit supplies temperature-controlled air at up to 36 cm H$_2$O (3.5 kPa) pressure. The air from the air blower is fed into a 1-kW heater unit, which is controlled by a digital temperature controller mounted on the control panel of the blower unit. The temperature is monitored by a sensor in the bed. The two-stage air blower is mounted on the extended shaft of a 0.75-HP (560-W) single-phase induction motor. The patient can control both contour and attitude positions, since the controls can be mounted on the side of the bed.

7.2.2 The Simpson–Edinburgh Low Pressure Air Bed

The Bio-engineering Unit at the Princess Margaret Rose Orthopaedic Hospital in Edinburgh developed the Simpson–Edinburgh Low Pressure Air Bed (figure 7.4). The bed consists of two standard camping air mattresses placed one on top of the other upon a wooden base with padded sides. Tubing connects the mattresses together and to an air pump. In the original design, the pressure of the bed was controlled by a discharge coil placed in a column of water. The air pump inflated the mattress to a preset pressure and when the patient was placed on the bed air was discharged to keep the pressure constant. The depth of the coil in the water regulated the pressure of the bed. Changes in posture were accommodated by more air being discharged or by a pause in the discharge of air until the original pressure was regained. The commercially available models of this bed feature a mechanical discharge valve instead of the discharge coil and water column. The air bed prevents the patient from bottoming out, while the adjustable pressure control ensures that the bed is not overinflated for the patient's weight.

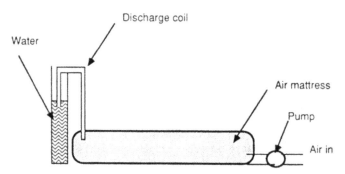

Figure 7.4 Principle of the Simpson–Edinburgh Low Pressure Air Bed.

Laing and Walton (1979) produced a study of the Simpson–Edinburgh Bed for the Scottish Home and Health Department. They concluded that the bed could reduce the frequency of turning necessary, but recognized some problems with the bed including instability, difficulty in changing the bottom sheet, difficulty in rolling the patient onto his or her side and problems when using bedpans.

7.2.3 The air fluidized bed

The air fluidized bed was first described by Hargest and Artz in 1969 for treatment of burns. Air fluidization is an easy way to achieve many of the desirable requirements of patient support with a mechanically simple system. Air is compressed by a blower and then forced through a diffuser and into a mass of granular material (figure 7.5). This mass consists of silicosises, soda lime glass spheres in a range of 75–100 μm. When the air passes up through the mass, the spheres separate from one another and are suspended in the airflow issuing from the blower. By such means, a "fluid" is generated. The fluidized mass has a density 1.45 times that of water; thus the patient floats upon the bed rather than in it. When in the static state, the material appears to be a fine white sand, but when fluidized it looks much like boiling milk (Hargest 1976).

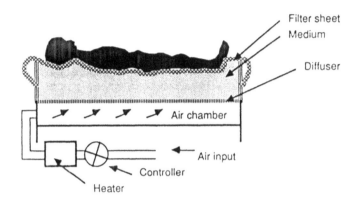

Figure 7.5 Air flows through small glass spheres in the air fluidized bed system.

The air fluidized bed (AFB) has a filter sheet placed over the mass. This sheet is of monofilament polyester and it does not absorb liquid. It can be washed in place if not badly soiled. It is woven in such a way as to have a tightly controlled 30 μm pore space. This prevents the 75–100 μm diameter beads from passing through it, but it permits the air to pass easily at a velocity of about 0.6 m/min. This is not high enough to be felt but is sufficient to cool anyone lying upon it when the bed temperature is less than 30°C. So normally the air is heated during the compression stage to about 39°C. Because of the large volume of glass spheres, rapid changes of temperature are difficult to achieve. Normally it takes about 2 h to heat such a bed from room temperature to 31°C, which is the lowest temperature at which the bed is normally used. If the bed becomes too warm it will cool at a rate of only about 1°C/h.

Any object immersed in the fluidized bed will be subjected to hydrostatic pressure for a given depth and density. Let us suppose a patient lies at a depth of 20 cm below the surface, then the pressure at that depth is

$$(20 \text{ cm } H_2O)(1.45)(0.7355 \text{ mm Hg/cm } H_2O) = 21.3 \text{ mm Hg.}$$

Pearce (1971) used a 2-cm^2 single bladder air sensor (section 8.3.2) to measure peak pressures. Average heel pressures from three male subjects reduced from 100 mm Hg on a regular bed to 40 mm Hg on an air fluidized bed.

The intermittent fluidized bed (IFB) has the same blower without heaters, heat exchanger, or filter sheet. This bed has a vinyl cover and is only fluidized for a few seconds every six to ten minutes, recontouring itself to the patient at such regular intervals.

The AFBs that breathe have important advantages compared with water beds. Water beds provide some shear forces and do not ventilate. One of the problems with AFBs is evaporative water loss. Patients must be protected by intravenous drips checked for adequacy by frequent serum sodium measurements.

AFBs have been used in the management of patients with pressure sores and severe burns in the institutional setting. They have been used both in hospitals and at home. Allman *et al* (1987) conducted a clinical trial to compare the effectiveness and adverse effects of air-fluidized beds and conventional therapy for patients with pressure sores. Their findings suggested that air-fluidized beds are more effective than conventional therapy, particularly for large pressure sores. Nimit (1989) proposed guidelines for home air-fluidized bed therapy. There is a growing trend toward and interest in the use of AFBs in the home as an alternative to hospitalization in the treatment of pressure sores. Home use may be optimized by applying patient selection criteria and a treatment protocol and by giving attention to those areas of potential complications that are preventable. The problems associated with use of AFBs, such as due to excessive heat caused by temperature control malfunction, are rare but have been reported. Spillage of beads, which may contaminate the environment and affect the skin and eyes, is also a risk that has been reported. More common problems include dehydration, dry skin, and alterations of mental status.

7.3 WATER BEDS

The water bed is a low-pressure support system which more evenly distributes the weight of a body over the total supporting surface. Therefore the pressure gradients and thus tissue distortion are reduced. To fulfil the criteria for true floatation a water bed must have sufficient volume to displace the patient's mass without developing tension in the cover. The cover must be flaccid, unlike the usual tough vinyl waterbed covers. Of the available water beds only the deep tank models can approximate true flotation. Other water beds may reduce tissue distortion to some extent, but as part of the patient's weight is supported by tension in the cover they cannot be considered as a preventive measure by themselves.

The Beaufort-Winchester Flotation bed (Paraglide Ltd) is one of the few support systems able to provide total hydrostatic support. The Beaufort-Winchester Bed has a depth of 38 cm. Shallow water beds (depth 16.5 cm) do not have the volume for true flotation but are less expensive and more mobile. However, it is doubtful if they are any more effective than some of the lighter inexpensive static support systems available. An intermediate water bed has a depth of 28 cm. Another type of water mattress is similar to camping air mattresses but is filled with water. Other systems use separate water chambers which can be assembled to yield a complete mattress. But these types of water beds have the disadvantages of weight and leakage without providing hydrostatic support. Deep-tank beds can provide hydrostatic support but the problems associated with these beds have included hypothermia, restriction of movement, difficulties in transferring the patient, instability, nausea, and disorientation.

When using a water bed, the mattress should only be partly filled with water so that for normal size patients flotation conditions can be achieved. The cover of the water mattress should remain in a deliberately loose condition after water has been added. The normal condition of use should be within a water temperature range of 30–35°C. The patient should never be placed on a bed at a temperature above 38°C. A temperature in excess of 42°C can cause skin damage. The bed should be used as a preventive device rather than for use solely for curing existing pressure sores.

Siegel *et al* (1973) measured the pressure distribution produced under the bony prominences of 10 volunteers on a water bed. The water bed used was Jobst Hydro Float® bed and a regular VA Hospital bed served as a control. The insulating liner tested was a Jobst Space Cloth®. A liner between the skin and the vinyl cover of the water bladder is necessary to prevent the moisture and skin maceration that direct contact produces. A space pressure manometer was used to measure the pressure. Figure 7.6 shows the results. When correctly filled, the water bed generally provided pressure of less than 20 mm Hg (capillary pressure). An exception was the calcaneal area (heel), where the mean pressure was 27.2 mm Hg. The data presented demonstrate that use of the water bed may be an important part of prophylaxis against pressure sores in paralyzed patients.

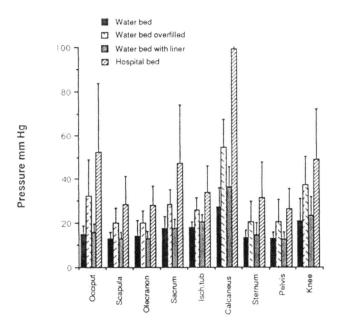

Figure 7.6 The water bed yielded lower pressures than a hospital bed (vertical thin line shows standard deviation)(Siegel *et al* 1973).

7.4 SAND BED

Dry sand is a clean, nearly sterile material with an adequately high density for "flotation" (silicon dioxide specific gravity 2.6). Based on this idea, Stewart (1976) used static sand beds on which the mattresses had been replaced by a tray of sand for preventing pressure sores. He also specified the sand and the bed. Dry sand does not cause an allergy, and silicosis or siliceous granuloma are dust diseases due to much smaller particles than those present in washed sands. Additionally it is capable of absorbing discharges and, being cheap, can be regarded as disposable. Sand beds provide the comfort experienced on a soft, dry beach.

In making a sand bed it should not be made flat. It is better to make manual impressions for the patient's sacrum, shoulders, and heels. The more immobile the patient the more "form fitting" impressions should be. Experience has shown that a sand depth of 12 cm will accommodate most contours (approximately 225 kg weight of sand). Sand is good at temperature maintenance so that there is minimal heat loss. A clinical thermometer placed in the sand an inch (25 mm) or so away from the body will show this. Stewart used a sand bed method for over five years. During this period no patient developed a pressure sore.

7.5 ASSISTED TURNING BEDS

Assisted turning beds have electrical motors to aid in turning the patient, thus saving nursing time. Fully automated turning beds are designed to turn the patient continuously from side to side to mimic the movements of a normal sleeping patient. Examples of these beds are the Egerton Turning and Tilting beds, the Stryker Circo-electric Bed, and Steeper Mini Co-Ro Treatment Table.

7.5.1 Automated bed

Roemer et al (1975) developed an automated bed to aid pulmonary drainage and prevent pressure sores for a 5-year-old girl. Before she used the bed, her parents manually turned her from side to side every 2 h to prevent prolonged pooling of fluid in one side of her lungs, and to prevent pressure sore formation. During this time she had been sleeping with her head-to-toe axis slightly inclined from the horizontal with head lowered to allow drainage of fluid from her lungs and out of the side of her mouth.

The automated bed (figure 7.7) replaced the manual rotation of the child from side to side with an automatic motor-driven motion, and, in addition, provided automatic and periodic tilting of her head-to-toe axis from the head-down position to the head-up position, thus providing more efficient pulmonary drainage and preventing fluid pooling and susceptibility to pneumonia.

While the system was originally designed to operate with a side-to-side delay (period between positional changes) of up to 2.5 min and an up-down delay of up to 12 min, the patient's parents observed that she sleeps much better and drains better with an up delay of 1 min, a down delay of 6 min, and a side-to-side delay of 30 min. The side-to-side period was much shorter than the usual period required to prevent pressure sores, but close to the period between movements of "naturally" sleeping subjects (average 11 min). The long head-down time allowed a sufficient period for drainage of fluid out of the lungs, along the throat, and out of the mouth, while only a short head-up period was needed to drain the upper lungs out of the shorter path into the lower lungs. Patients with different afflictions who have different time-delay requirements could use the adjustable-position limit and delay-period features of this design to meet their varied needs.

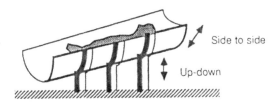

Figure 7.7 The automated bed tilted the patient from side to side to prevent pressure sores.

7.5.2 Bed turner

Hunter *et al* (1983) developed a mechanical bed called a Bed Turner. The bed was made for disabled persons who cannot obtain the services of a night attendant and required periodic turning during sleep. The turning is done mechanically and can be initiated by the patient or set to turn automatically. The Bed Turner mattress is divided and hinged longitudinally. It is suspended by pivots and bearings that allow the center board to tilt. A person sleeping on the bed with the mattress in a tilted position, perhaps to the right, would feel most of the mattress pressure on the right side of the body. After some time period, the mattress changes its position to a flat back position or to the opposite side position. This results in a shift in the body pressure areas. A true side laying position is not achieved, but there is a redistribution of pressure. This eliminates prolonged localized pressure areas, and as a result, prevents pressure sores. Control of the mattress position is done with electric circuitry. The patient can select the mattress position by operating the controller in its manual mode. The tilt of the bed can be changed by pressing a button on the control unit. In this mode, a timer causes the bed to turn at preset intervals. The interval between turns can be adjusted from 1.5 to 2.5 h.

Figure 7.8 shows a commercial Turning bed made by LIC Rehab Care company (Solna, Sweden).

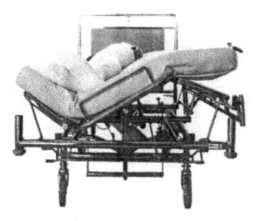

Figure 7.8 The LIC turnover bed has a center section that alternately inclines to the right or left.

7.6 NET SUSPENSION BED

Net suspension beds have been available to health authorities in the United Kingdom for a number of years. The Mecabed introduced by Mechanaids Ltd. is a net suspension bed which consists of a stainless steel framework supported by two plastic-covered steel cross members. Incorporated in the structure are two high-tensile steel poles, controlled by a clutch mechanism, which support a knot-free polyester net. The net requires no special laundering, and being polyester can be included with the hospital's normal white linen wash as it is not affected by detergents or bleaches. The Mecabed provides a smooth, knot-free surface, and to a large extent meets the requirements for distribution of local pressure. The net provides a strong support with sufficient elasticity which allows it to mold to the contours of the person's body, a property which does not appear to be lost with washing. Another net suspension bed is the Egerton net suspension bed (Egerton Hospital Equipment Ltd) which is a free-standing unit and can be turned from side to side to a maximum of 45°.

Brislen (1982) evaluated the Mecabed. The problem with using the Mecabed was that the tension which is applied to the net influences not only the pressure but also the patient's freedom to change position. If the net is pulled tight, there is a minimal restriction on the patient's body movement. When considering the adequacy of ventilation, a net covered by a single sheet provides the most efficient method of ensuring adequate ventilation. This benefit can be severely reduced if underblankets are used, and the use of bed pads for incontinent patients should be avoided as much as possible.

7.7 PRESSURE DISTRIBUTION ON DIFFERENT BEDS

One way to evaluate the effectiveness of different beds to prevent the development of pressure sores is to measure the pressure distribution on the body. Jeneid (1976) measured the pressure distribution on the LALBS and the King's Fund (KF) Bed. The KF bed (Ellison Ltd, Department of Health and Social Security contract model) is a bed with a longitudinally ribbed steel base supporting a 4-in (100-mm) polyether foam mattress, density 32–34 kg/m^3, supplied by Vitafoam Ltd, Middleton, Manchester, with a waterproof polyurethane-coated nylon cover. He used eight separate indicator pads which are described in Chapter 8 to measure the pressure at eight different locations. Figure 7.9 shows that with the exception of the head, elbow, and thigh values, the greatest range of pressures were observed on the KF bed when the patient was in the sitting position. The eight areas studied were the head, scapula, elbow, sacrum, buttock, thigh, calf, and heel. The maximal values recorded, in excess of 50 mm Hg, were the heel and the buttock on the KF bed, followed by the elbow on the LALBS. Most values show the ability of the LALBS to distribute the body load more evenly so that all parts of the body make maximal contact with the support surface. The KF bed has a much greater range of mean values in each of the three positions due to some areas taking little load and the remainder taking exceptionally high loads.

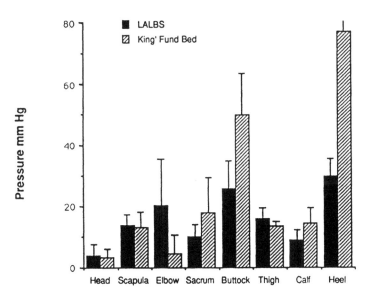

Figure 7.9 The LALBS yields lower pressures than the King's Fund Bed with the patient in a sitting position (vertical thin line indicates standard deviation)(Jeneid 1976).

Krebs *et al* (1984) measured pressure distributions on two different air flotation beds, the Clinitron bed and the Mediscus bed. The Mediscus bed is comprised of air-inflated pillows that permit a controled amount of air to escape from them. The pillows are connected to five manifolds so that the stiffness of the support surface can be adjusted independently under the head, thorax, abdominal area, legs, and feet. The Clinitron bed uses an air-fluidized bead support system that provides a uniform stiffness over the entire support surface. The pressure measurement were made using a Texas Interface Pressure Evaluator (TIPE). Subjects were tested while lying in the supine and side lying positions. The bony areas that were monitored during the data-collection sections were the scapulae, sacral-coccygeal region, and trochanters. The results of this investigation were based on an analysis of the maximal pressure measured under the bony prominences on each of these beds. The data showed that the mean values of pressure on the Mediscus were between 19 and 24 mm Hg and on the Clinitron were between 19 and 25 mm Hg among thin, normal, and heavy subjects. An analysis of the statistical significance of the differences in the means was conducted using the Student T test. The results of this analysis showed that there was no statistical significance.

Krouskop *et al* (1985) evaluated the pressure reduction characteristics of seven mattress surfaces. Thirty subjects were selected and evaluated using the Texas Interface Pressure Evaluator (TIPE). Seven support surfaces used were the Standard hospital mattress, Stryker flotation system, Roho mattress, 2-inch (50-mm) Convoluted foam, 4-in (100-mm) Convoluted foam, Akros DFD support system, Gaymar alternating air mattress and Lapidus alternating air mattress. Figure 7.10 shows the pressure under the scapula while subjects lay in

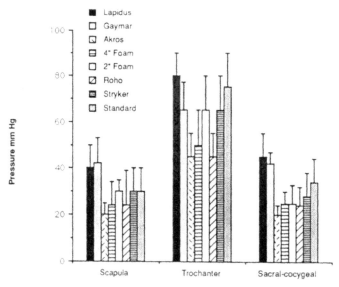

Figure 7.10 Pressure under three locations while subjects lay in the supine position on seven different mattress surfaces (vertical thin line shows standard deviation)(Krouskop 1985).

a supine position on seven different mattress surfaces. A greater difference was observed for the Roho and Akros overlays. Under the sacrum there was also a significant difference between the overlays and the standard mattress. They also measured pressure under trochanter and sacral-coccygeal region.

Ryan *et al* (1989) measured pressure at seven body points in four commonly available beds: the Standard bed, Water bed, Low air loss bed, and Fluidized bed. They used a small water-filled sensor attached to a transducer to measure the pressure. Figure 7.11 shows the data. A lower pressure was measured with all three beds over the occiput and the heel. The water bed showed a more even distribution of pressure than the standard bed. The low air loss bed and the fluidized bed were notable for the very low pressures they achieved, except for the occipital region with the low air loss bed.

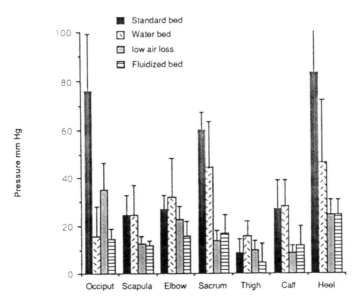

Figure 7.11 The fluidized bed yielded the lowest pressures for supine patients (vertical thin line shows standard deviation) (Ryan 1989).

7.8 AUTOMATED WHEELCHAIR SEATS

Chapter 6 presents various wheelchair designs. One potential method of alleviating the effects of the high pressure is to periodically remove pressure from the tuberosities by mechanical methods, as is done in various alternating air cushion mattresses. Houle (1969) discussed this approach which emulates the natural reaction of unimpaired individuals who relieve pressure through movement, and that of patients who perform "wheelchair pushups". Tally Medical Industries has marketed a pneumatic system for alternating pressure, Roemer (1979) developed an oscillating wheelchair, and Kosiak (1976) introduced a system based on a series of moving rollers that provides periods of pressure relief.

7.8.1 Oscillating wheelchair seat

Roemer (1979) developed an oscillating wheelchair seat for the prevention of pressure sores based on the ideal of cyclically lowering the surface underneath the ischial tuberosities to provide periodic pressure relief.

Figure 7.12 is an exploded view of the major seat components. The complete system easily fits into a wheelchair. For installation, the normal wheelchair seat is removed and replaced by the oscillating seat with a cushion placed on top. The cushion has a cutout area of the appropriate size, normally located under the area of the ischial tuberosities. Lowering of the seat is accomplished passively by a return spring. Raising of the seat to the level position is accomplished by a cable, pulley, and spring system that is driven by a reversible permanent-magnet gear motor through a worm gear reduction system. Power is obtained from two 6-V rechargeable nickel cadmium batteries with a life of 1 Ah. Based on an operation cycle of raised 30 min, lowered 3 min, the system can operate for 50 h before discharging the batteries completely. Thus, with nightly battery charging there is no danger of discharging the system during normal operations.

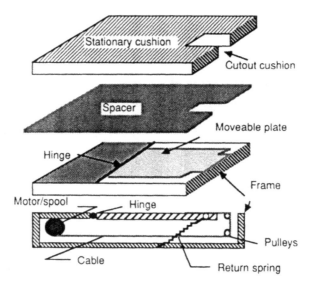

Figure 7.12 In the oscillating wheelchair seat, a spring passively lowers a hinged moveable plate and a motor-driven cable raises it to the level position (Roemer 1979).

Roemer (1979) measured pressure and temperature change during operation of the seat under the ischial tuberosities. Pressures were measured with

Scimedics Inc. 2-mm-diameter pneumatic pressure sensors, and temperatures were measured using iron-constantan thermocouples taped directly to the skin adjacent to the ischial tuberosities. The experiments showed that during the times when the seat was lowered, pressure dropped from 45 mm Hg to 25 mm Hg. Then, as the subject gradually sank lower into the cutout area and the cushion in the cutout area expanded, the tuberosity pressure rose to a higher steady state value which was about 35 mm Hg. After 2 h operation, the temperature rose about 0.5°C which indicated that an increased blood perfusion of the tissues was taking place, thus providing the tissues with nutrients and removing metabolic waste products. He emphasized that it was this process that should help in the prevention of pressure sores.

7.8.2 Mechanical resting surface

Kosiak (1976) noted that a 72-kg person has an upper body mass of 54 kg (529 N force). The available sitting area is 0.06 m^2. A uniform pressure would be 529 N/0.06 m^2 = 8.8 kPa (66 mm Hg). He developed a mechanical resting surface to help redistribute pressure. The seating area of a wheelchair was replaced by a mechanical resting surface which consisted of a series of rollers operating on a continuous belt assembly. This was powered by a small direct current motor, which was energized by four D-size rechargeable batteries. The entire belt assembly moved in an anterior-to-posterior direction, with each freely rotating, supporting roller moving approximately 1 cm/min or each roller displacing the adjacent one every 6 min.

   He also measured pressure under six areas of the buttocks using CT-200+ sensors (Scimedics, Inc., Paramount, CA). Beneath the ischial tuberosities pressure extremes ranged from 0 to 160 mm Hg depending on the position of the supporting roller devices. This device provides the sitting patient with complete relief of pressure at regular, predictable, 3-min intervals. That is, one half of the sitting area at all times is completely free of pressure for up to 3 min. Since the supporting rollers are in constant motion, the body is subjected to maximal pressure for only short periods of time. The pressure distribution curve produced by the resting surface resembles a sine wave, showing a maximal pressure only momentarily before deceasing to near-zero pressure, then rising again as the next roller support moves into position. Though short periods of high pressure are not desirable, these pressures may not be as destructive as low pressures (above capillary pressure) when applied for prolonged periods of time. However, the project was abandoned.

7.8.3 Weight transfer wheelchair seat

Joyner (1988) designed a weight transfer wheelchair seat which provides pressure relief to the buttocks area of the wheelchair user who is unable to shift his or her own weight to prevent pressure sores. The seat can be used with a solid state timer developed for the purpose or can be operated manually (figure 7.13).

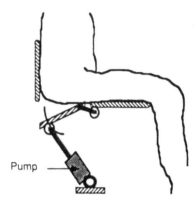

Figure 7.13 Weight transfer wheelchair seat. As the hinged part lowers, it provides pressure relief.

7.9 STUDY QUESTIONS

7.1   Sketch the block diagram of a controller for an alternating pressure mattress.
7.2   Explain why the low air loss bed is superior to a simple air mattress.
7.3   Explain the advantages of the air fluidized bed.
7.4   Sketch the pressure profile on a body on a waterbed with a compliant cover.
7.5   Sketch the pressure profile on a body on a waterbed with a normal cover.
7.6   Explain how a sand bed can cause flotation.
7.7   Sketch an automated bed and show extreme motions.
7.8   Describe the operation of the Bed Turner.
7.9   Explain how the net suspension bed can prevent pressure sores.
7.10 In a comparison of LALBS and King's Fund beds, which subject position produced the largest pressure and at what location for each?
7.11 Of seven mattress surfaces tested, which two were most effective?
7.12 Sketch and discuss the use of the oscillating wheelchair.
7.13 Sketch and explain the operation of the mechanical resting surface.

8

Bladder Pressure Sensors

Ren Zhou

This chapter presents the bladder-type sensors which readily conform to the curvature of human soft tissue and the support media, have good repeatability in the measurements, are inherently insensitive to shear forces and temperature changes and thus they are widely used in the measurement of pressure at the skin–cushion interface. We introduce general design criteria of sensors and explain the principles of different bladder sensors. In the last section, we evaluate them based on results from a simulation test.

8.1 SENSOR SPECIFICATION

8.1.1 Ideal sensor

An ideal sensor system would not perturb any parameters at the body/support interfaces but measure the true interface pressures. The sensor system would measure and monitor pressures from 0 to 1500 mm Hg, cover the area of interest, have zero diameter or very small sensing areas that resolve the pressure gradient (figure 8.1), have zero thickness and larger compliance than body tissue or the support media, have no hysteresis, not introduce nor respond to humidity and temperature, and not produce any artifact.

8.1.2 Sensor diameter

We can estimate the maximal diameter of a sensor by giving the distribution of pressure or pressure gradient and the measurement error limit (figure 8.1). Assume that the sensor is of uniform sensitivity, responds directly proportionally to the mean pressure applied to its area, and has radius r'. Assume the spatial distribution of the pressure at the interface to be measured is isotropically distributed and directly proportional to its distance from the center of the distribution, i.e. $P(r) = P_0 - kr$ where k is the pressure gradient, P_0 is the peak pressure at the center of the distribution and r is the distance from the center. By integrating the pressure over the sensing area, calculating the mean

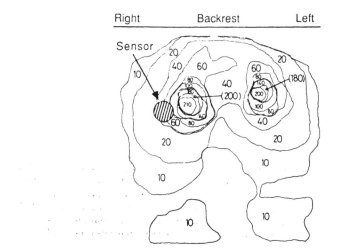

Figure 8.1 A typical pressure gradient map of seated position with the shaded sensing area overriding different pressure contours (Ferguson-Pell 1980). The maximal sensor radius can be estimated with an assumed error limit and the pressure gradient. The numbers indicate pressures in mm Hg.

pressure and comparing it with the peak pressure under the given error limit, we can solve for the radius of the sensor (Ferguson-Pell 1980)

$$r' = 3P_0e/200k \qquad (8.1)$$

where e is the measurement error allowed.

Applying equation (8.1) to a few practical cases with an error limit of 5% and known values of k and P_0, Ferguson-Pell suggested, based on the averaged result, that the maximal diameter of a sensor be 14 mm. In order to resolve the pressure distribution as shown in figure 8.1, a total number of 10,000 sensors with a diameter of less than 4 mm in an area of 40 × 40 cm would be required to give an accuracy within 5% of the peak pressure.

8.1.3 Sensor thickness

Ferguson-Pell (1980) also estimated the maximal thickness of a sensor by giving the material constants, dimensions, the applied pressure and the error allowance. Assuming that both the human soft tissue and the support material are linear, elastic, homogeneous, isotropic and not time dependent, then the estimated thickness h is given by

$$h = \frac{wlPe}{2}\left(\frac{1}{k_sl + 2a_st_s} + \frac{1}{k_ml + 2a_mt_m}\right) \qquad (8.2)$$

where w and l are the width and the length, respectively, of a rectangular sensor; P is the applied pressure; e is the measurement error; and k, t, and a are expressions relating the material constants of the soft tissue or support medium when they are denoted by the subscripts s or m, respectively:

$$k = \frac{Ew}{H(1 - v^2)}\left[\frac{H}{2l}\left(\frac{\sinh H/l \cosh H/l + H/l}{\sinh^2 H/l}\right)\right] \qquad (8.3)$$

$$t = \frac{E}{12(1 + v)}\left[\frac{3l}{2H}\left(\frac{\sinh H/l \cosh H/l - H/l}{\sinh^2 H/l}\right)\right] \qquad (8.4)$$

$$a = (k/2t)^{1/2}. \qquad (8.5)$$

In equations (8.3), (8.4) and (8.5), H is the thickness of the material, E is compressive stress–strain modulus and v is Poisson's ratio of either human tissue or support medium when they are denoted by the subscripts s or m, respectively.

Example 8.1 Given $w = 10$ mm, $l = 10$ mm, $E_s = 5$ kPa, $v_s = 0.25$, $H_s = 10$ mm, $E_m = 25$ kPa, $v_m = 0.15$, $H_m = 100$ mm, $P = 100$ mm Hg (13.3 kPa) and $e < 5\%$. Substituting the numbers into equations (8.2), (8.3), (8.4), and (8.5), we have the maximal thickness of sensor $h < 0.03$ mm.

8.2 SPATIAL DISTRIBUTION

8.2.1 Seated

Figure 8.2 shows a typical sensor arrangement. This is useful in studying the pressure distribution covering the whole seating area. Some researchers put a single sensor under one ischial tuberosity provided the sensor size is large enough and only the average pressure of the area is of interest (Palmieri *et al* 1980). We can also individually tape small sensors to the skin at spots of interest.

8.2.2 Recumbent

Ideally we would use a sensor pad that covers the whole upper surface of the bed when studying pressure distribution for any recumbent position. Then the subject can move or be moved about in the bed. But if only a few special areas are of interest, we can arrange the sensors based on information from studies on overall pressure distribution (figure 1.2 and figure 6.14). An example of a recumbent sensor distribution is the bed-of-nails (figure 10.1) where sensors are spaced every 1 cm.

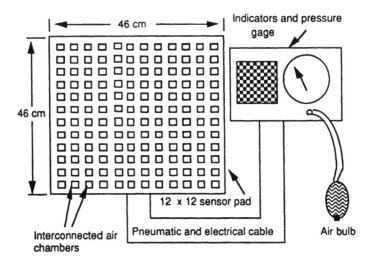

Figure 8.2 The Texas Interface Pressure Evaluator (TIPE) pad measures pressure with 144 plastic air bladders.

8.3 BLADDER SENSORS WITH BUILT-IN SWITCH

8.3.1 Principle of operation

A bladder sensor with built-in switch, as shown in figures 8.3(b) and (c), consists of an air bladder, a tube to let air in and out, a pair of electrical contacts (as a switch) fixed to the opposite inner sides of the bladder, an inflation/deflation device, an air pressure gage and an electrical indicator wired to the switch.

During a test, the air bladder is placed at the skin–cushion interface. Before inflation, the electrical contacts in the bladder touch each other and thus close the circuit as monitored by the indicator. Then the inflation device pumps in the air, and the pressure in the bladder increases. The bladder as well as the electrical switch become open when the air pressure is equal to or greater than the external pressure at the skin–cushion interface. At the moment the switch opens, the electrical circuit opens, and the pressure gage reads the value which is supposed to be the interface pressure.

The sensing system can also work the other way around. The pump first inflates the air bladder to a pressure greater than the external pressure so that the switch opens. Then the pump deflates the bladder. At the moment the switch closes as monitored by the indicator, the pressure gage reads the value of the interface pressure.

Introducing any practical sensor(s), including the air bladder(s), at the interface can change the interface pressure (figure 8.3, section 8.1, and section 8.5).

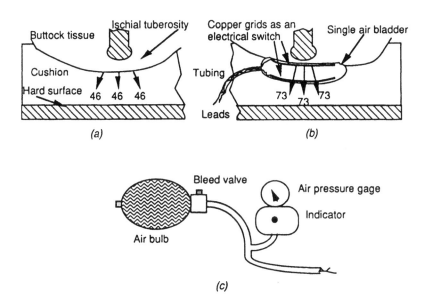

Figure 8.3 *(a)* Undisturbed skin–cushion interface with arrows and numbers indicating the interface pressure in mm Hg from a simulation test (Reger and Chung 1985). *(b)* Cross section (not to scale) of a single air bladder sensor which increased the interface pressure to 73 mm Hg in the simulation test (section 8.5). *(c)* The device to inflate and deflate the air bladder and to read the pressure at the moment the switch opens or closes.

8.3.2 Single bladder air sensor

Pearce (1971) used a 2-cm^2 single bladder air sensor to measure peak pressures. Average heel pressures from three male subjects reduced from 100 mm Hg on a regular bed to 40 mm Hg on an air fluidized bed.

Figures 8.3(*b*) and (*c*) illustrate a typical product from Scimedics, Inc. It is a 9 × 10 cm oval, 0.5 mm thick inflatable plastic bladder with copper grids attached to its inner surfaces as electrical contacts. The pump–gage system consists of a hand-operated rubber bulb, an aneroid manometer, and plastic tubings that carry the air and the electrical wires.

The lower and upper pressure limits of the sensor are 0 and 120 mm Hg, respectively (McGovern and Reger 1987). The pressure values under ischial tuberosities are often as high as 200 mm Hg (figure 8.1). However the Scimedics bladder measures the average pressure applied over its wide area, so the values are usually lass than 100 mm Hg (Mayo-Smith 1980). The data deviations in repeated tests are within 17% at about 26.5 mm Hg and 4.3% at about 92 mm Hg.

The drawback of this sensor is its large size. We can not use it to resolve fine pressure contours like the ones in figure 8.1. Also, one-sided application under the ischial tuberosity may cause a subject's weight shifting and imbalance which a healthy subject can feel. This changes the pressure distribution and results in measurement error.

8.3.3 Multisensor bladder pad

Figure 8.2 shows the exterior of Texas Interface Pressure Evaluator (TIPE) pad. It consists of separate top and bottom sheets of translucent vinyl plastic bonded together around the outside edge (Jaros *et al* 1986). It is also quilted together across the face to form 144 small interconnected air chambers. These chambers are arranged in a square 12 × 12 matrix and lie at 32-mm intervals. The deflated thickness of the pad is about 1 mm. The pad is air-tight and can be inflated to a desired pressure through an attached hose. Twelve thin silver strips are painted on to the inner sides of each sheet (figure 8.4). Strips on one sheet run horizontally to form rows and those on the other sheet run vertically to form columns. A row crosses over a column at the center of each air chamber. A small rectangle of resistive material is arranged at the intersection so that an electrical switch connection through this material is formed between the row and column whenever the top and bottom surfaces physically contact. A given switch closes when the externally applied pressure exceeds the internal pressure of the pad at that point.

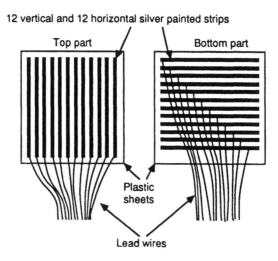

Figure 8.4 External pressure on the Texas Interface Pressure Evaluator pad causes the vertical and horizontal silver painted strips to contact.

A microprocessor-based switch scanner/pressure monitor can be designed to interface the TIPE pad with a personal computer so that the operation of the system is quick and easy (Chapter 10).

The weakness of the TIPE is that it acts as an air cushion if it is used with a support surface of low compliance. It is likely to reduce the applied pressure in adjacent high-pressure areas, leading to inaccurate readings (Bader and Hawken 1986).

8.4 BLADDER SENSORS WITHOUT BUILT-IN SWITCH

8.4.1 Henry surface pressure manometer

The Henry surface pressure manometer (figure 8.5) consists of an air bladder made of latex, a pair of tubes, A and B of the same volume, a pair of mercury globules in the tubes, a rubber inflation bulb, a buffer tank and a mercury manometer. The bladder is fixed at the end of tube A, tube B is closed-ended.

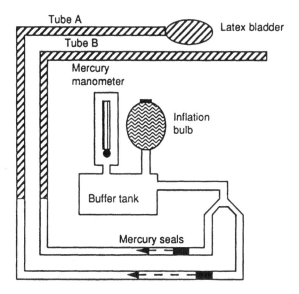

Figure 8.5 Henry surface pressure manometer (Souther 1974). The lower mercury globule advances to the left of the upper globule when the latex bladder is inflated open at the skin–cushion interface and then the manometer reads the interface pressure.

To measure the pressure, the latex bladder is placed at the skin–cushion interface and the air bulb inflates the bladder via the buffer tank and tube A. When the external pressure at the interface is greater than the internal pressure, the air pushes both mercury globules leftward at the same speed. At the moment the internal pressure equals the external pressure, the bladder opens up, and the system volume through tube A becomes larger than that of tube B. So the lower globule in tube A passes to the left of the upper one in tube B. This indicates the moment when the manometer reading should be recorded as the interface pressure. The upper limit of the system is 130 mm Hg.

8.4.2 Oxford air bladder system

Based on the principle of the Denne gage (Crewe 1983), which is a simple pneumatic device composed of an inflated compliant bladder connected to a mercury column, Bader *et al* (1985) developed the Oxford Pressure Monitor (figure 8.6). It consists of a 3 × 4 bladder matrix which is placed at the skin–

cushion interface, a 12-to-1 airway selector, a pressure polarity controller, a pump, a semiconductor strain gage sensor, numeric and analog displays, and a single chip microprocessor as the system controller.

Individual air bladders, 20 mm in diameter, are spaced at distances of 28 mm between centers. The deflated thickness of the pad is about 1 mm. Each air bladder, with a hose connected to the selector, is independent of the others in the pad.

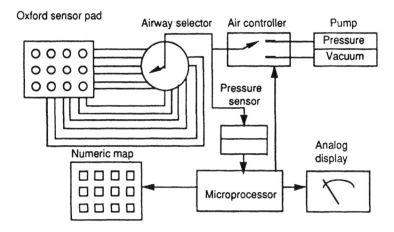

Figure 8.6 Oxford Pressure Monitor (Bader and Hawken 1986). The pump injects a constant air flow to a selected bladder and the microprocessor detects the change of pressure gradient (temporal) in the selected bladder when it opens.

At the start of the measurement cycle the pneumatic bladders are held in a deflated condition by the vacuum side of the pump (Bader and Hawken 1986). Next, the air controller is switched to the pressure side of a high-pressure pump, which provides a constant mass flow via a needle valve to inflate the selected bladder. At the point when the pressure in the bladder becomes equal to the external pressure, the cell begins to open. This produces a change in the volume of the system, thereby reducing the rate at which the internal pressure increases (figure 8.7). The pressure gradient (or the rate of change) is continuously monitored by the microprocessor by means of the pressure sensor. The system pressure at the instant the pressure gradient changes is recorded, and the air controller is immediately switched back to vacuum to deflate the bladder. The measurement cycle is shorter than 0.5 s for one bladder at 250 mm Hg which is the upper limit of the system.

While measuring the pressure applied to the selected bladder, the other bladders not in use remain deflated. They do not distort the local pressure distribution (Bader and Hawken 1986). But we presume that the weight shifting effect as with the Scimedics single air bladder (section 8.3.2) may still contribute to measurement error. Also, because of the distances between each of the 12 bladders, the spatial resolution may not be fine enough.

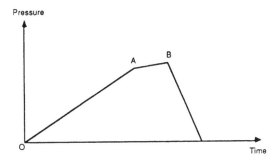

Figure 8.7 The internal pressure of a bladder in the Oxford system increases at a constant gradient from point O to point A while the external interface pressure is larger than the internal. When they become equal at point A, the bladder opens, the volume increases, the gradient changes and is detected. The pressure at point A is then recorded as the interface pressure. The pump starts to suck out the air from point B.

8.4.3 Rubber butterfly valve

Kosiak *et al* (1958) used 12 flat rubber butterfly valves, 2 cm long, 1 cm wide, and located 4 cm apart, as the sensors under the ischial tuberosities (figure 8.8). The valves were attached to a nondistensible closed system into which a steady flow of air was maintained from a pressure source of approximately 600 mm Hg. Each of the 12 valves is calibrated individually at various pressures up to 300 mm Hg against a Hunter Force Indicator.

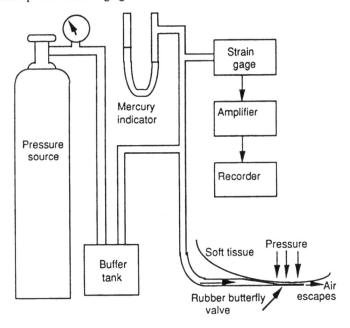

Figure 8.8 The subject sits on a rubber butterfly valve. The internal pressure increases until the constant flow air escapes (Kosiak *et al* 1958).

1180118   PREVENTION OF PRESSURE SORES

During a test the valves are connected in succession to a strain gage system
by means of stopcocks. The pressure at which equilibrium is reached between
inflow of air into the system and escape of air through the test valve is accepted
as the pressure applied externally on the valve.

8.4.4 Water bladder system

A water bladder system (Holley *et al* 1979) consists of ten water-filled, water-
tight, air-tight bladders, each connected via vinyl tubing to a 10-to-1 waterway
selector which leads to a diaphragm-type pressure sensor, the electrical output
of which goes to an amplifier with analog display and to a digital voltmeter for
better resolution (figure 8.9).

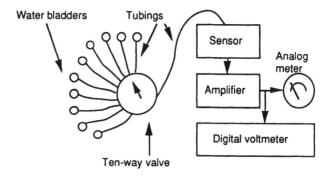

Figure 8.9 A water bladder system uses water to transmit pressures (Holley *et al* 1979).

The water bladder consists of a vinyl membrane clamped over a metal ring,
10 mm internal diameter, 17 mm outside diameter and 5 mm thick. When a
force is exerted on the bladder the pressure developed is transmitted to the
sensor which converts the water volume displacement to an electrical signal
with a sensitivity of 0.5 μV/mm Hg. The signal is then amplified and the
voltage displayed.

Ryan and Byrne (1989) developed a small plastic ellipsoid bladder sensor,
made of two circular walls sandwiched together. It forms a vessel that can hold
a small volume of water, less than 2 mm in thickness and with a contact area of
2.25 cm^2. It is attached to a Shaevitz pressure sensor and the total system
displaces less than 5 ml of water. The sensors are filled with proper amount of
water so that they remain flexible, not producing stresses on the walls which
would lead to a nonlinear response. They calibrated the system by applying a
known force on the bladder and used it to evaluate four common hospital beds.

8.5 EVALUATION

Reger and Chung (1985) evaluated the Scimedics single air bladder, the Texas
Interface Pressure Evaluator pad, and the Oxford bladder system in three
simulation setups (figure 8.10).

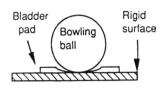

Maximum peak pressure(mm Hg) Hertz contact stress: 47,000			
Load per area on sensor pad	TIPE	Scimedics	Oxford
6,800	200	138	68

(a)

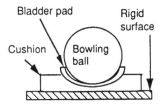

Mean contact pressure(mm Hg) Hertz contact stress: 45.8			
Load per area on sensor pad	TIPE	Scimedics	Oxford
68.3	54.1+1.0	73.3+1.1	92.3±0.1

(b)

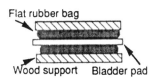

Applied pressure (mm Hg)	Measured pressure (mm Hg)		
	TIPE	Scimedics	Oxford
20	20+1	20	15.6+1.0
40	40+1	40	39.2+1.0
60	60+1	60	60.8+1.1
80	80+1	80	81.1+1.8
100	100+1	101	102.0+1.4
120	120+1	122	122.3+1.0
140	140+1	141	141.4+1.8
160	160+1	162	159.5+1.7

(c)

Figure 8.10 Simulation setups to evaluate Scimedics, TIPE and Oxford pressure sensors (Reger and Chung 1985). (*a*) Bladder(s) between a bowling ball and hard surface; (*b*) bladder(s) between a bowling ball and a cushion; (*c*) bladder(s) between plane rubber bags.

They first used Hertz methods (Burr 1981) to calculate the theoretical contact stress for each case in figure 8.10 without the bladder pad at the bowling ball–support interface. Hertz methods provide solutions for deformation, area of contact, pressure distribution, and stresses at the initial point of contact of two elastic bodies with convex surfaces or one convex and one plane surface or one convex and one concave surface. Then they placed different bladder pads one at a time at the interface and measured the pressure for each case (*a*), (*b*) and (*c*) shown in figure 8.10.

For the case in figure 8.10(*a*), bladder(s) between a bowling ball and hard surface, the result turned out to be the worst. The measured values were at least two orders of magnitude smaller than the theoretical one. They claimed that the location of the contact spot relative to the center of the sensor was most critical with the TIPE pad and least with the Scimedics bladder.

For the case in figure 8.10(*b*), bladder(s) between a bowling ball and cushion, they found that the measured values were greater than the theoretical value (also figure 8.3) because all three types of bladder pads stiffen the ball–cushion interface. For multibladder pads, TIPE and Oxford, the surface tension or Hammock effect (Chapter 5) can also increase the pressure. However, Bader and Hawken (1986) suggested that TIPE, as a thin cushion, is likely to reduce the pressure, probably when the compliance of the support surface is relatively low. For example, when TIPE is between a hard surface and ischial tuberosities, the interface pressure will be reduced. In figure 8.10(b), TIPE showed the least error in absolute pressure values while Oxford presented the smallest deviation of values in repeated tests.

For the case in figure 8.10(c), bladder(s) between plane rubber bags which are inflated to apply the pressure, the result was almost ideal. The measured values for all three types of bladder pads agreed very well with the known pressures applied.

Note that in figures 8.10(a) and (b), they used the bowling ball, a rigid object, instead of a soft object to simulate human tissue. A bowling ball wrapped with a soft material would have been a closer simulation to find the theoretical contact stress, which is beyond the scope of the Hertz formula.

Researchers in the area of body/support interface pressure measurement have claimed that available sensors, including the bladders, are useful in relative pressure comparison at different locations of the same subject and the same cushion or among different subjects and/or different support surfaces. Bladder sensors have presented better measurement repeatability than other types of sensors.

8.6 STUDY QUESTIONS

8.1 List the specifications of an ideal interface pressure sensor.

8.2 List the specifications of a practical interface pressure sensor.

8.3 Explain the constraints in determining the diameter of a sensor.

8.4 In order to ensure the measurement of the pressure under the ischial tuberosities, how many sensors should be placed over what area?

8.5 In order to ensure the measurement of pressure under the trochanter, how many sensors should be placed over what area?

8.6 Carefully sketch a Scimedics bladder sensor under an ischial tuberosity. With arrows show all pressures and forces. Explain why it measures correct pressure or any errors in the measurement.

8.7 Repeat the previous question for the Texas Interface Pressure Evaluator pad (TIPE).

8.8 Explain the principle of the Henry surface pressure manometer.

8.9 Sketch the pressure-flow curve and explain the operation of the Oxford Pressure Monitor.

8.10 Sketch and explain the principle of the rubber butterfly valve sensor.

8.11 For a water-filled bladder pressure sensor, explain how the volume is maintained and the effect of volume changes on accuracy.

8.12 How did Ferguson-Pell determine the dimensions of a pressure sensor as being 14 × 5 mm?

8.13 Explain how a sensor can interfere with or change the pressure distribution that is present when the sensor is not there.

9

Conventional Pressure Sensors

Jesse Olson

Chapter 8 presented methods of measuring interface pressure which use devices designed specifically for that purpose. In this chapter we present various types of conventional pressure sensors which have applications for interface pressure measurement. These include conductive polymer sensors, semiconductor and metal strain gages, capacitive, and optoelectronic sensors. We also present several types of pressure sensors for making measurements of internal pressures.

9.1 GENERAL CONSIDERATIONS

Chapter 8 describes the many constraints placed on sensors used for interface pressure measurement in order to disturb the "true" interface pressure distribution as little as possible. Ferguson-Pell (1980) calculated that the overall sensor thickness should be less than 0.5 mm, and the diameter should be less than 14 mm. To measure spatial distribution of pressure to within 5%, sensor diameter should be less than 4 mm.

In addition to these concerns, there are special considerations that become significant when measuring interface pressure with conventional sensors.

9.1.1 Force vs pressure

We are interested in measuring pressures, because it is pressure that collapses vessels and causes pressure sores. We may use a force sensor if we know the sensing area, because pressure equals force divided by area. A force sensor will respond identically to two equal forces regardless of the area over which the force is applied. A true pressure sensor will, under the same constant force conditions, give an output inversely proportional to the area. Therefore, in order to use a force sensor in a pressure-sensing application, care must be taken to ensure that the force is always applied over the same area. Strain-gage sensors actually sense strain, but can be used to indirectly measure pressure. Both absolute and differential strain-gage pressure sensors exist. Most sensors measure the average pressure over the sensing area, but to measure pressure distributions we want to measure peak pressures. As the sensing area decreases, the assumption of uniform pressure over the sensing area becomes more valid.

Examples of force sensors include some optoelectronic sensors. Most capacitive sensors are true pressure sensors, as are fiberoptic catheter sensors with pressure-sensitive membranes. Conductive polymer sensors are pressure sensors if the load is applied uniformly over the entire sensing area.

The effects of off-axis loading must also be considered. If a sensor has high directional sensitivity, it may not respond accurately when a flexible mat material conforms to an irregular surface. Most sensors are designed to respond to forces normal to a sensitive face, but off-axis loads also contain shear forces. Some capacitive sensors are sensitive to shear forces, as are microbending fiberoptic sensors and conductive polymer sensors. It is important to be able to isolate the measurement of shear forces from the measurement of normal forces. When applying sensors which are sensitive to flexing, care must be taken to ensure that no curvature occurs in the area local to the sensor.

9.1.2 Deformable materials

All conventional pressure sensors depend on the deformation of a material to generate a pressure-dependent change. There are several key characteristics of deformable materials which influence the operating characteristics of the sensor. These include linearity, hysteresis (memory), creep (drift), and yield. Linearity is preferred, but nonlinear outputs can generally be overcome by computerized calibration methods. Measurement repeatability will be significantly reduced if the output of a sensor depends on whether pressure is increasing or decreasing. For interface pressure measurements, a dynamic range with a time constant on the order of a few seconds is sufficient. If the applied forces cause the material to exceed its elastic limit, its response will be permanently altered.

Pressure-sensitive materials include elastomers, silicon, and metal. The deformation characteristics of these materials vary greatly. Elastomers, for example, have significant nonlinearity, hysteresis, and creep but can withstand large overloads. Silicon, although it is fragile, has a very high elastic limit, and negligible hysteresis and creep. It is, however, inherently nonlinear. Various metals exhibit a range of characteristics. Beryllium copper is an example of a metal with high elasticity and low hysteresis, and is often used as a spring material.

9.1.3 Nonfluid medium

Most conventional pressure sensors have been designed to measure fluid pressures. However, human–seat interface pressure measurements are generally made by placing the sensor between two layers of compliant solid material such as silicone rubber. Therefore, some sensors may require modification or careful application to produce meaningful results. Patterson and Fisher (1979) pointed out some of the difficulties with using conventional sensors in interface pressure measurement. Also, the deformation characteristics of encapsulants and mat material must be considered since they may significantly affect sensor performance.

The investigator must be aware of whether a given sensor is measuring peak pressure or average pressure. Errors may result if an assumption of uniform pressure over the entire sensor area is incorrect.

Care must be taken also that the mat material or sensors do not cause the "hammock effect" mentioned in Chapter 5, and so distort the "true" pressure distribution. Sensors that are too large or mat material that is not sufficiently compliant may cause this problem.

9.1.4 Measurement range

For interface pressure measurements, the maximal pressure we will need to measure will be about 500 mm Hg for a subject seated on a hard surface (Ferguson-Pell *et al* 1976). Therefore, any sensor we use must be able to withstand pressures in that range without yield or fatigue. Fortunately this is a relatively low pressure, so most sensors can withstand this range. The temperature range may vary from ambient (20°C) to body temperature (37°C), while humidity may vary over the entire range. If sensor response varies as a function of temperature, temperature data must be taken along with pressure data, and the calibration protocol must take temperature variations into account.

In order to accurately measure a pressure distribution, a large number of sensor locations will be necessary. This implies the need to multiplex many sensors to some recording or central processing location. The difficulty of multiplexing many sensors depends on the nature of the output of the sensor, and whether or not pre-amplification is required. Large resistance changes or low impedance DC voltages are easily multiplexed, as are digital outputs. Very small resistance or capacitance changes are not as easily handled. With fiberoptics many outputs can use a common detector if some method of switching between sources exists.

9.2 CONDUCTIVE POLYMER PRESSURE SENSORS

9.2.1 Force-sensing resistor

Most conductive polymer pressure sensors yield a decrease in resistance with increasing pressure. This change is the result of an increased contact area between two sheets of polymer (or elastomer), such as Mylar or Ultem, coated or impregnated with conductive material. Figure 9.1 shows the construction of the Interlink force-sensing resistor, or FSR. The top sheet is coated with an elastomer containing metal powder or carbon black. The bottom sheet has a silk-screened pattern of interdigitated silver conductors.

The conductive portions of the two sheets are in minimal contact in an unloaded state, and the resistance between the two lands on the bottom sheet is normally infinite. The two sheets may not be in contact at all, but separated by a thin (10–150 μm) spacer around the outside of the conductors. With a small load, the resistance decreases to a value from 1 to 10 MΩ, with a final resistance of about 1 kΩ. The "turn-on" pressure depends on the presence and/or thickness of the spacer. The relationship between pressure and resistance is nonlinear. This type of sensor is referred to as shunt mode design, since the resistance between the two connections is shunted by the top sheet. A second mode of operation is through-conduction, in which the two sheets have conductive areas similar to that of the top sheet in figure 9.1. Tests have shown that through-conduction is not as stable as shunt-conduction at higher forces.

Establishing a connection to the conductive pattern poses some problem. The elastomers from which these sensors are made have relatively low melting points, so conventional hot solder techniques cannot be used. Possible solutions include crimp connectors, silver epoxy, low-temperature solder, and nickel paint. Strain relief for the lead wires should be provided such as by heat shrink tubing. Interlink manufactures the sensors with three connection possibilities: crimp-connected terminals which may be soldered, two-pin connector plugs, or with plain conductive lands as shown in figure 9.1.

A simple voltage-divider circuit can be used to convert resistance to voltage. By placing the force-sensitive resistor in the upper half of a voltage-divider circuit, an increase in voltage will correspond to an increase in pressure. A second connection method places the sensor in series with a current source. An advantage of this method is that the operating range of the sensor can be easily controlled by adjusting the current source.

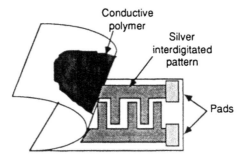

Figure 9.1 The no-load resistance of the Interlink force-sensing resistor is infinite. As pressure forces the two sheets together, the silver interdigitated pattern is shunted by the conductive polymer of the top sheet (Interlink 1987).

9.2.2 Performance characteristics

The repeatability of a given sensor is highly dependent upon the geometrical relationship between the sensor and the load. Although the sensor is flexible, it should not be allowed to flex during measurements because the response curve may change dramatically. The sensor is typically attached to a coin-shaped flat metal backing. If the alignment between load and sensor is carefully maintained, repeatability can be as good as 1% (Interlink 1987). From sensor to sensor, the repeatability is about 20%, which means that each sensor must be calibrated individually. To increase the repeatability, the upper surface of the sensor should be covered with a thin layer of compliant material, such as 1.5 mm silicone rubber. This tends to apply the load to the sensor more uniformly. Loads should be applied normal to the measuring surface, since operation in the presence of shear forces may be unpredictable.

To compensate for nonlinearity and differences in individual sensors, one of several calibration methods must be employed. These include static and dynamic methods, and may make use of regression or look-up tables to interpolate between calibration points. Chapter 11 presents these methods in greater detail.

The electrically active ingredients consist of metallic type conductors with a positive temperature coefficient of resistance combined with semiconducting components which have a negative coefficient. The result is a relatively low temperature coefficient, on the order of $-15\ \Omega/°C$, which corresponds to about $1\%/°C$.

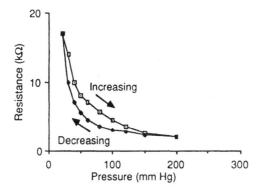

Figure 9.2 An Interlink sensor calibration curve exhibits hysteresis. The arrows indicate increasing and decreasing pressure.

9.2.3 Interface pressure sensing

There are several qualities of conductive polymer sensors which make them suitable for use in interface pressure measurement. They are very thin, typically on the order of 250–400 μm. Since these sensors are fabricated using screen printing techniques, they can be made in a wide variety of shapes and sizes, or be fabricated into an array. They are inexpensive, with costs under $1 in quantity. To use a conductive polymer pressure sensor as a pressure sensor, care must be taken to ensure that the load is applied uniformly over the entire sensing surface.

Because the conductive polymer sensors exhibit large changes in resistance, they are relatively easy to use. Resistance is easily converted to voltage, and the large changes make more complex bridge circuits and high-gain amplifiers unnecessary. Multiplexing of these sensors could be accomplished in a straightforward manner.

Due to the excellent resilience of rubber, conductive polymer sensors can withstand large overloads. They can be subjected to many compression cycles before material fatigue becomes significant.

However, there are several disadvantages of these sensors. The performance of these sensors is nonlinear and nonrepeatable. Although they respond to dynamic as well as static loading, the elastic properties of the polymer sheets exhibit significant hysteresis and creep, so they are not well-suited for very accurate or long-term pressure measurements. Figure 9.2 shows hysteresis typical of a conductive elastomer sensor. It can be seen that repeatability of measurements will suffer since a given measurement depends on whether pressure is increasing or decreasing. They do not necessarily return to the same no-load value. When a static load is applied, the resistance of the sensor may continue to decrease for minutes. In fact, the time constant of the creep characteristic is such that the time from loading to measurement must be carefully observed. The suitability of these sensors for long-term measurements is questionable.

Piche *et al* (1988) proposed the use of an array of 0.25-mm-thick foil resistors (similar in design to the force-sensing resistors described earlier) to measure human–seat interface pressure. Since the grid was so thin, the elastic properties of the interface were almost unaltered during measurements. A multichannel data-acquisition system was used to sample the pressures, and voltages were converted to interface pressure by a numerical curve fit algorithm.

Maalej *et al* (1987, 1988) used Interlink sensors to measure pressures under the foot. Many of the criteria for sensors to measure foot–sole interface pressures are similar to those for measuring pressure at the buttock–cushion interface. A multichannel ADC system was employed to sample seven sensors at 20 Hz. They found that the best connection method was to attach multistrand wire to the sensor pads with silver epoxy after abrading the Mylar with sandpaper.

Pax *et al* (1989) used Interlink sensors to measure palmar pressures. They coated 32-gage stranded wires with silicone rubber to provide strength and flexibility, then attached them with silver paint. They encapsulated the sensor in RTV silicone rubber to provide strain relief for the connections and a more uniform pressure distribution. The RTV layer was then sandwiched between two thin sheets of silicone elastomer. The package had an overall thickness of 2 mm. They explored several methods, but found that a plain sensor responded more repeatably and over a larger area than sensors with a single backing plate or both a top plate and a backing plate. However, in their application the sensor was always applied against a smooth, hard surface, so it was not subject to flexing during measurements.

Webster (1989) proposed the design of a mat which contained an array of conductive polymer pressure sensors. Preliminary interface pressure measurements with a single Interlink sensor for a hard seat and for a foam cushion showed dramatic differences, indicating promise for the use of an array of these sensors in a thin mat. We have constructed such an 8 × 8 array of discrete 0.018" thick Interlink sensors (Part #174, 7/8" circle plain sensors) sandwiched between layers. The base was a 0.025 × 12 × 12" pure gum mat. Sensors were glued on top of 0.025" aluminum metal backups to minimize hysteresis. Belden 8430 audio wire was snaked between the sensors. Silicone rubber filled any voids to smooth out discontinuities. Transparent plastic formed the top layer. Figure 10.14 shows the resulting real-time computer-generated three-dimensional plot generated by an IBM PC.

There are many problems with using the mat as proposed as an alarm device, because of the individual differences in susceptibility to development of pressure sores. However, it might find use in measuring pressure distributions in the design of support surfaces.

9.3 SEMICONDUCTOR STRAIN GAGES

9.3.1 Piezoresistivity

Semiconductor strain gages operate on the piezoresistive principle. In pure silicon or germanium, a valence electron can be excited into the conduction band to participate in the conduction of electric current. A "hole" in the crystal lattice is left behind, which can also serve as a current carrier. The excitation voltage required to free carriers is about 1 eV in pure ("intrinsic") silicon, which accounts for the relatively high resistivity of intrinsic semiconductors.

To decrease the resistivity, foreign atoms are added by "doping" the material with elements of the third or fifth group of the periodic table such as boron or phosphorus. Because phosphorus has five valence electrons to the four of silicon, the fifth electron is only weakly attached and may be excited to the conduction band with only 0.05 eV. This is referred to as negative- or n-doping. If boron is used as a doping agent, a fourth bond with silicon is weak and only 0.08 eV is required to fill the hole with an electron from an adjacent silicon atom. This is referred to as positive- or p-doping.

The piezoresistive effect, the variation of resistance with strain in strips of silicon or germanium, is primarily due to variations of the carrier mobilities in these materials. For semiconductor strain gages the sensitivity is given by the gage factor

$$F = \frac{dR/R}{dL/L} = 1 + 2v + m.$$

The lateral contraction term v (Poisson's ratio), which accounts for dimensional changes in the material, is about 0.5 and can be neglected for semiconductor strain gages in comparison to the change in resistivity m. This term is the product of the piezoresistive coefficient and Young's modulus, and is typically around +175 for p-type silicon, or −130 for n-type silicon. This compares with a typical gage factor of +2 for metal strain gages.

Semiconductor strain gages have many desirable characteristics. Silicon sensors are predictable and reproducible. Although silicon is fragile, it has an elastic limit greater than steel. It can be stressed repeatedly without weakening, and retains measurement accuracy. Hysteresis and creep effects are negligible. If these effects are observed in an application, they are probably due to the material used to attach the sensor element. Semiconductor strain gages do have nonlinear response curves, but the amount of nonlinearity is relatively small. However, they are significantly temperature dependent, so some method of reducing temperature dependency must be incorporated. The amount of doping affects the temperature dependency as well as the sensitivity, so a reasonable compromise must be made. Various compensation techniques can be employed. Two- or four-active arm Wheatstone bridges are often used. The bridge is useful for measuring small resistance changes, and complimentary arrangements of strain gages tend to reduce temperature dependency.

Most semiconductor strain gage sensors are made by diffusing resistors on the surface of a thin diaphragm. The diaphragm, typically formed by electrochemical etching, is on the order of 10 μm thick and 500 μm in diameter (Esashi *et al* 1982). As pressure is applied, the diaphragm deflects and the resistance values change accordingly. Diaphragm thickness directly affects the sensitivity of the sensor. Four resistors are typically used in a Wheatstone bridge arrangement. The transverse voltage sensor, developed by Motorola, is a unique four-terminal device which uses a single resistive element, eliminating the need for closely matched resistors (Derrington *et al* 1982). Guckel's pill box is similar to a diaphragm sensor, but lateral etching techniques are employed for very small diaphragm sizes on the order of 100 μm by 100 μm (Guckel and Burns 1984).

There is a wide range of semiconductor strain gage sensors offering various levels of temperature and nonlinearity compensation. An example of a high-end sensor is the Kulite Semiconductor Products CQ–030 series at a cost (1987) of $892. This sensor uses a four-active-arm Wheatstone bridge, and is 0.76 mm in diameter by 3.75 mm in length. It is fully compensated and ready to use. A typical low-cost sensor is the Sensym SCX C series at $15, but it is only useful for measuring gas pressures. Strain gages are available which include only the resistive elements, in packaged or die form. Attachment of a strain-gage element to a surface may cost $100, including wire bonding of leads.

9.3.2 Interface pressure measurements

Most semiconductor strain-gage sensors have been designed to measure fluid pressures. When applying these sensors to the measurement of interface pressure, certain difficulties arise. Figure 9.3 shows a typical semiconductor diaphragm pressure sensor. Resistors are diffused into the top surface of the diaphragm. The small port in the housing of the IC strain gage is not sufficient to transmit pressure in compliant materials to the diaphragm, so sensitivity is

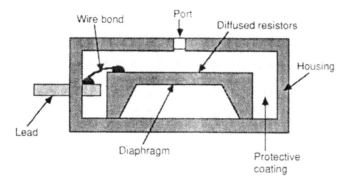

Figure 9.3 A typical semiconductor diaphragm pressure sensor for use in a fluid medium. The resistive elements are diffused into the diaphragm, usually in a Wheatstone bridge configuration (Sensym 1988).

very low. But if the housing material is removed, the fragile wire bond is only protected by the coating. When subjected to flexing, the wire bond will break after only a few repetitions.

In order to solve this problem, Beebe *et al* (1990) explored a new packaging technique, as shown in figure 9.4. Starting with an unpackaged diaphragm sensor mounted temporarily on a Teflon disk, they applied several layers of polyimide. They then etched the polyimide to reveal the bonding pads on the diaphragm, and deposited thin film metal leads. After applying more polyimide layers to protect the leads, they removed the Teflon, and replaced it with a backing disk. The backing disk had the effect of increasing the contact area of applied force. They applied final layers of polyimide to yield a sensor which was totally encased. A similar technique has been used successfully in the packaging of temperature sensors. The output of the sensor would vary with the compliance of the tissue, so calibration procedures should include placing the sensor between layers of material with compliance similar to that of tissue. The main problem with this approach was the inability to planarize the die–substrate interface via spin coating. Another problem was the warping of the Teflon substrate at relatively low processing temperatures. Thus they are fabricating their own pressure sensors using micromachining. Silicon etching using KOH etches shallow (< 5 μm) cavities so the diaphragm will bottom out during overload. Silicon fusion bonding will bond the constraint and sensing wafers. Ion implantation will create a piezoresistive Wheatstone bridge in the diaphragm. Polyimide layers will encapsulate the sensor.

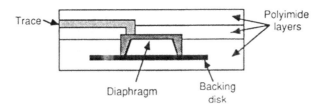

Figure 9.4 The polyimide layers protect the connections to the IC, eliminating the fragile wire bond of figure 9.3.

To measure pressures at the foot–shoe interface, Soames *et al* (1982) placed a single semiconductor strain gage on a uniquely designed cantilever of beryllium copper, as shown in figure 9.5. The entire sensor was 0.9 mm thick and 13 mm

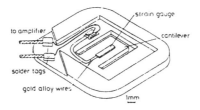

Figure 9.5 Pressure deflects the cantilever into the recess. Displacement is measured by the attached strain gage (Soames *et al* 1982).

square. To accurately measure pressure, the surrounding material must be sufficiently compliant to distort the cantilever without significantly changing the pressure. Also, the load must be evenly distributed over the cantilever. The sensor was calibrated by applying a known pressure through a compliant membrane. The calibration data for this sensor were quite linear up to 160 kPa. Special care must be taken when applying a sensor of this type to ensure that the compliance of the material used in calibration is similar to that of the material (such as tissue) being measured.

Because silicon sensor fabrication uses the same technology as that of integrated circuit manufacturing, electronic circuits can be incorporated into the design of the pressure sensor. This permits several functions such as temperature sensors, signal conditioning, compensation, and calibration to be incorporated into a single package. This may be very useful in interface pressure measurement, since temperature has been suggested to play a role in the formation of pressure sores. Also, since arrays of sensors imply long lead lengths to a central processing location, preamplification and/or digitization may be useful in reducing crosstalk and noise.

Semiconductor sensors may be as small as $1 \times 1 \times 0.5$ mm (Nunn and Angell 1975). Because of their small size and compatibility with body fluids, silicon sensors are often used in implantable applications to measure internal pressures. These applications are discussed in Section 9.7.3.

9.4 METAL STRAIN GAGES

9.4.1 Design theory

Metal strain gages, unlike semiconductor strain gages, exhibit a change in resistance which is primarily due to dimensional changes. As an axial force is applied to a bar of metal, the bar becomes longer and its cross-sectional area decreases. Strain ε is the change in length per unit length, and Poisson's ratio v is defined as lateral strain per longitudinal strain. From the equation for resistance of a metal bar, we can develop the following expression for the gage factor

$$F = 1 + 2v + \frac{d\rho/\rho}{dL/L}$$

where ρ is the resistivity of the metal, and L is the length. The gage factor expresses the dependence of resistance change on strain. For most metals in their elastic region, v is about 0.3. Most metals have a gage factor of about 2 (Hagner 1988).

Most metal strain gages are made from alloys of nickel and copper. Metal strain gages have been manufactured in a wide variety of styles, with attempts to optimize the performance characteristics by creating multiple turns or lengths in a single gage.

9.4.2 Performance

The dimensional changes and piezoresistive effect are very small, so the gage factor of metal strain gages is significantly less than that of semiconductor devices. An advantage of metal strain gages is that their response is nearly linear, although they have a significant temperature dependence. A Wheatstone bridge is generally used with metal strain gages to sense small changes in resistance. The bridge also helps reduce problems due to temperature changes, but matching the resistance values precisely can be expensive.

Chizek *et al* (1985) used four Precision Measurement 105 pressure sensors to measure pressures at the foot–shoe interface. These sensors are 280 μm thick and have a diameter of 2.7 mm. Although these sensors are supplied ready to use, in order to optimize mechanical load coupling Chizek *et al* packaged them in Silastic MDX elastomer (silicone rubber), with the sensor resting on a thin rigid sole. This had the effect of equalizing pressures applied directly over irregularities in the surface of the sensor package. The cost of this sensor is about $175.

At the University of Wisconsin, we have built two cantilever sensors, similar to that shown in figure 9.5, using metal strain gages. One cantilever was made of spring steel, and the other of beryllium copper. The gages had a base resistance of 120 Ω and a full scale deflection of 0.5 Ω. We found that the sensors were extremely sensitive to the compliance of the material used to transmit the load to the cantilever.

9.5 CAPACITIVE PRESSURE SENSORS

9.5.1 Capacitance variations

A capacitor is a charge storage device in which the voltage across the capacitor depends on the amount of charge stored. The equation for capacitance is

$$C = \frac{\varepsilon A}{d}$$

where ε is the dielectric constant, A is the effective plate area, and d is the separation between the plates. A capacitive sensor can measure pressure by changing any of these parameters.

Capacitive sensors in general are very sensitive and fairly linear. Discrete capacitive sensors may suffer from tradeoffs between physical size and capacitance values. If capacitance values are too low, the circuitry necessary to detect even smaller changes becomes complex, and shielded leads must be used to reduce problems caused by stray capacitance. Unfortunately, larger capacitance values imply increased physical size. The cost of capacitive sensors is dependent upon the process used to manufacture them.

Since capacitance is not readily observable, it must somehow be converted to a measurable quantity, typically voltage. Frequency-to-voltage circuits are often used. Another commonly used circuit sends a fixed-charge pulse to generate a voltage proportional to capacitance. The impedance bridge can be used to convert small capacitance variations to DC voltages (Seow 1988). These circuits are significantly more complex than the voltage divider or bridge amplifier circuits discussed in previous sections.

9.5.2 Interface pressure sensors

A flexible sensor designed by Neuman et al (1984) was used to measure forces at finger and thumb tips. The capacitor consisted of foil or metalized plastic film plates separated by a dielectric composed of a mixture of silicone oil and silicone elastomer. Figure 9.6 shows that the dielectric contained air channels to increase its apparent compliance. Thick film screen printing or molding can be used to manufacture the dielectric. As pressure increases, the dielectric fills the air channels and the plates move closer together.

The sensors were fabricated both as discrete elements and arrays. The discrete sensors were 2 cm square and from 0.24 to 1.5 mm thick. Sensitivity and linearity were related to the composition of the elastomer–oil dielectric. Linearity was maximized for low oil-to-elastomer ratios, but sensitivity increased at higher ratios. Drift and hysteresis were greater in the thick film dielectric than the molded dielectric. However, larger capacitance values were obtained for the thick film dielectric. It may be possible to manufacture this type of sensor at relatively low cost in mass quantity.

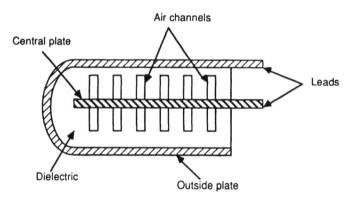

Figure 9.6 As force is applied to the capacitor plates, the dielectric fills the air channels and the plates move closer together (Neuman et al 1984).

Patel et al (1989) used a capacitive pressure sensor designed by Hercules (model #F4-4R) to measure pressures between the foot and shoe. Figure 9.7 shows that the sensor is constructed with four layers of mica as a dielectric. The dielectric is sandwiched between five plates of corrugated metal. Applied pressure flattens the corrugations, and the capacitance increases. This sensor has a diameter of 17 mm and a thickness of 2.4 mm. The capacitance changed from 275 to 580 pF over a pressure range from 0 to 1300 kPa. Sensitivity is greatest at lower pressures. The response of the sensor is nearly linear up to 500 kPa. A phase-locked loop was used to convert capacitance changes to voltage.

The sensor exhibited hysteresis on the order of 10%, and a temperature coefficient of –0.147%/°C. Nonrepeatability was about 7%. The cost of the sensor is $150. The sensitivity to point loads is not uniform and decreases from 1.0 at the center to 0.17 at the edge.

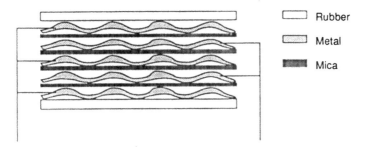

Figure 9.7 As pressure increases, the corrugations in the metal plates flatten out, increasing the capacitance of the sensor.

Ferguson-Pell *et al* (1976) discuss the use of capacitive sensors composed of a thin film of polyester adhesive. The sensors have a no-load capacitance of about 200 pF, but the full scale change in capacitance is only 5 pF. This type of sensor can be manufactured from inexpensive materials, with a thickness of approximately 400 μm and a sensitive area of 1 cm^2.

Figure 9.8 shows an example of monolithic silicon IC technology used to manufacture air-gap capacitors with signal processing circuitry on-chip. Although there are problems in reaching all of the potential advantages of capacitive sensors over piezoresistive types, IC capacitive sensors can be more sensitive, less temperature dependent, and smaller than their piezoresistive (strain gage) counterparts (Ko *et al* 1982, Smith *et al* 1986, Lee and Wise 1982). This type of sensor finds applications in implantable sensors, since it can be very small and the silicon from which it is manufactured is biocompatible.

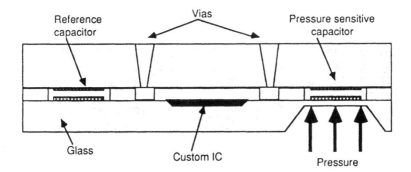

Figure 9.8 The design of this IC capacitive pressure sensor includes a reference capacitor and signal processing circuitry on-chip. Physical dimensions of the packaged IC are 6 mm × 3 mm × 0.5 mm (Smith *et al* 1986).

However, only fluids can reach the recessed diaphragm and it is not suitable for measuring interface pressure where fluids are not present.

Babbs *et al* (1990) present the most promising capacitance-sensing pressure-sensitive mat. Figure 9.9 shows that it uses two orthogonal arrays of ribbon-like conductors, composed of silver-coated nylon fabric. These are separated by insulating, open-cell, low-hysteresis foam rubber. The system monitors the capacitance between selected pairs of horizontal and vertical conductors on opposite sides of the foam. The crossing points form pressure-sensitive nodes. Increased contact pressure compresses the foam, thereby increasing the distance between the conductors and increasing the capacitance. Node capacitance is determined by measuring the current through it from a 5-kHz voltage source. Cross-talk through unselected shunt capacitances is eliminated by connecting unselected drive lines and sense lines to ground. Data are displayed on a 24 × 64-node array, as shown in figure 10.11.

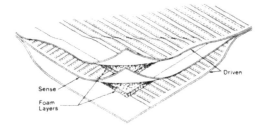

Figure 9.9 Five-layer construction of the pressure-sensitive mat. Three conducting layers are separated by two compressible foam insulators. Parallel sense lines are included in the middle conducting layer. Drive lines are included in the top and bottom conducting layers, which also act as shields. Pressure-sensitive nodes include two capacitances mechanically in series but electrically in parallel (Babbs *et al* 1990).

9.6 OPTOELECTRONIC PRESSURE SENSORS

9.6.1 Optoelectronic technology

Although many types of optoelectronic pressure sensors exist, the criteria necessary for interface pressure measurement have typically prohibited the use of these sensors. Optoelectronic sensors have generally been too thick, and costs have been high. However, the rapid development of optoelectronic technology, and fiberoptics in particular, will likely produce a suitable system in the coming years.

In an array of sensors, the advantages of using fiberoptics to transmit the signals stand out. Since fibers may be as small as 50 μm, spatial sensitivity can be very high if the sensor itself can be kept small. Fibers lend themselves to multiplexing techniques, reducing the number of signal-conditioning circuits necessary for an array of sensors. Fragility of fibers is a problem that must be considered, especially if they are to be placed in compliant mat materials. Also, coupling light into small fibers is a complicated task which may require expensive laser diodes or lens systems. Because of their low cost and ease of use, light-emitting diodes (LEDs) should be used whenever possible. For interfacing applications larger diameter fibers (1000 μm) would satisfy the spatial sensitivity requirements, and the optical coupling can be much less complex. We will explore some of the potential types of optoelectronic pressure sensing systems.

9.6.2 Occluder types

One of the most straightforward types of optoelectronic sensors places a movable optical barrier between source and detector as shown in figure 9.10. Sensors of this type include the Lord tactile sensing array (Dario and DeRossi 1985), and a discrete element sensor designed by Maalej and Webster (1988). The most significant problem with occluder types has been the size. The Lord sensor had a thickness of 6 mm, not including membrane material. The discrete element sensor designed by Maalej and Webster used separate source and detector elements, and had a thickness of about 3 mm. The sensor had relatively low hysteresis and good repeatability. Encapsulation in silicone rubber improved the linearity of the sensor but sacrificed sensitivity. The encapsulation cracked after 1000 cycles.

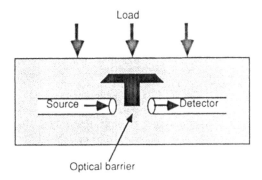

Figure 9.10 As pressure increases, the optical barrier blocks a greater portion of light.

By using a single package (TRW OPM101) containing both LED and phototransistor, Hahm and Webster (1989) designed a sensor, shown in figure 9.11, with a thickness of 1.9 mm. With a smaller optoelectronic package, the overall thickness of the package could be reduced. The sensitivity to force decreases from the right edge to the left edge. By placing the sensor between two layers of compliant material and attaching some hard disks of known area, its response to pressure could be improved. The sensitivity to pressure would vary with the compliance of the tissue, so calibration should include placing a material of similar compliance to tissue in series with the sensor. The physical size of these occluder type sensors could be reduced by the use of fiberoptics.

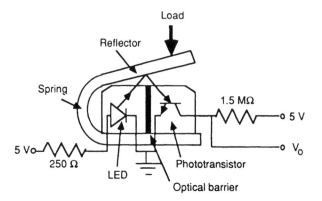

Figure 9.11 The LED and phototransistor are contained in a single package, TRW OPM101.

9.6.3 Intrinsic and extrinsic fiberoptic sensors

A fiberoptic sensor which uses deformation of the fiber itself as the pressure sensitive change is referred to as an intrinsic sensor. Figure 9.12 shows a type of intrinsic fiberoptic sensor, referred to as a microbending sensor. As pressure increases the channels cause irregularities in the core–cladding (jacket) interface of the fiber which increase transmission losses, also referred to as induced bending loss (Giallorenzi *et al* 1982). This type of sensor was used to measure pressure between adjacent teeth (Jamsa *et al* 1990). The microbending pattern was etched into $20 \times 11 \times 0.1$ mm steel plates. The fibers used were 50 μm in diameter. The total thickness of the sensor was 0.2 mm. This type of sensor is very sensitive to off-axis loading.

Herga Corporation produced a sensor which consisted of a spiral of flexible plastic wound around a fiber. As the pressure on the spiral increased, microbending patterns were produced in the fiber. This sensor was used in pressure-sensing mats and contact-sensing bumpers (Maalej and Webster 1988, Hahm and Webster 1989).

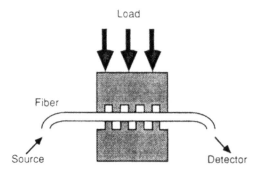

Figure 9.12 As pressure increases, the grooves cause bending in the fiber, and transmission losses increase proportionally (Brienza *et al* 1989a).

Fiberoptic sensors in which the fibers are used strictly for transmission of light to the sensing location are referred to as extrinsic sensors. One such type is a diffuse reflective sensor in which an optical fiber guides light from a source to a sensor. At the sensor, the light leaves the fiber and reflects off a diffuse surface to a second fiber. The second fiber guides the light to a detector element. The distance from the fibers to the reflecting surface is regulated by a linear spring element. The intensity of received light is then proportional to the applied pressure. While the use of optoelectronics and fiberoptics is attractive, these sensors are difficult to construct and have not been useful for interface pressure measurements.

In general, intrinsic fiberoptic sensors are of questionable use for long-term studies. A jacket must be used since the glass fiber itself is extremely fragile, so the deformation characteristics of the jacket material become significant, introducing hysteresis and fatigue into the sensor response. Additionally, since the strain of the measurement is placed on the fiber itself, the optical characteristics of the fiber will become degraded after numerous repetitions. Brienza *et al* (1989b) recommended diffuse reflective fiberoptic sensors over microbending sensors for use in automated seating design systems.

9.7 MEASURING INTERNAL PRESSURE

9.7.1 Needle manometer method

Because some pressure sores develop due to high internal pressure, it is desirable to know the pressure of the internal tissue. Internal pressures can be three to five times greater than external pressure (Barth *et al* 1984). There are few methods of determining this pressure *in vivo* in a simple, accurate, and reliable manner. A needle manometer can be used, but requires a continuous injection of heparinized saline to keep the catheter tip open. Without the continuous injection of fluid, the catheter tip may become clogged by particles or tissue walls. However, the injection disturbs the equilibrium conditions in the area of the measurement. Other disadvantages of the catheter method include the fact that the tissue pressure cannot be measured during muscle contractions, and long-term determinations are not possible.

9.7.2 Wick catheter method

The Wick catheter method is similar to the needle manometer method, but a wick of cotton or Dexon fibers protrudes from the catheter tip. Figure 9.13 shows that these fibers prevent obstruction of the needle opening by objects in the interstitial fluid or contact with tissue walls, and provide an extensive contact area for fluid equilibrium. Braided wicks composed of 10 μm diameter fibers are tied together with a monofilament suture. The wicks are pulled about 10 mm into a 1 mm (OD) catheter, leaving about 7.5 mm protruding. The suture is then threaded about 20 cm up the catheter and tied off. This prevents loss of the fibers inside the body. The catheter is then filled with heparinized saline and connected to an external sensor for measurement. The wick catheter is inserted into the body via a 16-gage catheter.

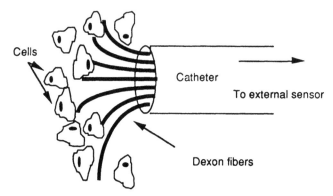

Figure 9.13 Wick catheter fibers permeate a large volume of interstitial fluid, and prevent blockage (Mubarak *et al* 1976).

Mubarak *et al* (1976) compared the wick catheter technique with the needle manometer method and a solid-state pressure probe. They determined that the wick technique was the most accurate and reproducible, likely due to the relatively large volume of tissue fluid permeated by the wick. They also successfully used the wick catheter technique to record interstitial pressure during muscle contractions.

By placing a wick catheter at very shallow depths in the tissue and measuring external pressure at the same location with a Scimedics air-cell sensor, Reddy *et al* (1984) found that the two methods agreed to within about 15% for various loading conditions.

9.7.3 Integrated circuit implantable sensors

There are various types of IC pressure sensors, including piezoresistive and capacitive types discussed in earlier sections. These sensors can be manufactured small enough to be placed in catheter tips or sidewalls, or implanted in the tissue for long-term pressure measuring in research studies. Implantable sensors do not suffer from losses due to hydrostatic coupling. Electrical safety considerations must be carefully maintained when using implantable ICs.

9.7.4 Packaging

Packaging of such sensors must be as biocompatible as possible. It must be noninflammatory, nontoxic, nonallergenic, and noncarcinogenic. The material must be mechanically durable and chemically stable. Silicon crystal is a good material for biosensors because of its biocompatibility (Hynecek 1975, Ko *et al* 1979). Good choices for wires include Elgiloy (a nickel–cobalt alloy), stainless steel, platinum–iridium, copper, or nickel. For insulation and/or encapsulation, Teflon, silicone rubber, and titanium have been found to have desirable qualities (Regnault and Piccolo 1987). Ko *et al* (1979) used Hysol W-795 epoxy for attachment of components and Silastic silicon rubber for encapsulation.

9.7.5 Fiberoptic catheter

Many of the packaging problems considered above could be solved by application of fiberoptic sensing elements. Figure 9.14 shows a simple fiberoptic catheter used to measure blood pressure. A thin (20 μm), metallized membrane is placed at the end of a fiberoptic catheter with a small air space (50–100 μm) between the ends of the fibers and the membrane (Christensen 1980). Changes in pressure cause deflections in the membrane, modulating the light intensity reflected to the detector fibers. The individual source and detector fibers may be interspersed within a single bundle. Proper selection of the equilibrium position of the membrane will result in a fairly linear response to pressure.

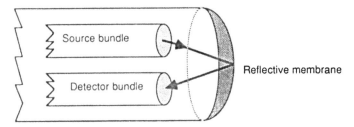

Figure 9.14 Pressure changes bend the reflective membrane and modulate the light received by the detector bundle.

This type of sensor has been used to measure blood pressure, and may require some adaptation for use in tissue. The fiberoptic method has advantages in that all the electronic circuitry can be placed outside the body. The risk of shock is removed, and losses due to hydrostatic coupling are eliminated. Glass fibers are highly biocompatible. Disadvantages of fiberoptic catheters include higher cost and complexity, and reduced reliability.

9.8 SUMMARY

Conventional pressure sensors offer many potential benefits over bladder-type systems for interface pressure measurement. Among these are improved accuracy and repeatability, increased spatial resolution, ease of use and lower cost. However, there are significant problems involved in realizing these advantages. The rigid constraints on thickness have been a major obstacle in implementing many types of conventional pressure sensors, while excessive hysteresis and drift have prevented the use of others.

Figure 9.15 compares the various types of conventional pressure sensors. For interface pressure measurements, there are other factors which are also of importance. The sensors should lend themselves to being placed in an array, and preferably multiplexed with a minimum of signal-conditioning circuitry. Considering cost and size as the most important factors, conductive polymer sensors are an excellent choice for an interface pressure measuring system. The sensors can be designed for size and shape as necessary, and arrays of sensors would be easy to multiplex. The resistive output is simple to measure or convert to voltage. The sensors are also very thin, which is critical for interface pressure measurement. Smaller sensing areas than those currently available may be desirable to spatially resolve the pressure distribution.

External types	Cost	Size (mm) W×L×H	Repeat-ability	Linearity	Hysteresis	Temp. coeff.	Ease of use
Conductive polymer	$1	10×15×0.2	Fair	Poor	Poor	Good	Excellent
Semicond. strain gage	$10–1000	1.0×1.3×0.4	Excellent	Good	Excellent	Fair	Good
Metal strain gage	$100–500	2.7×2.7×0.28	Excellent	Excellent	Excellent	Poor	Fair
Discrete capacitive	$50–200	15×15×0.4	Good	Good	Good	Good	Poor
Extrinsic optoelec.	$2–500	3.0×4.0×2.0	Good	Good	Good	Good	Fair
Intrinsic optoelec.	NA	20×11×0.2	Good	Good	Good	Good	Poor
Internal types							
IC capacitive	$50-1000	2.0×3.0×0.4	Excellent	Good	Excellent	Excellent	Good
IC semicond.	$50-1000	1.0×1.3×0.4	Excellent	Good	Excellent	Fair	Good
Wick catheter	$10	1.0 OD	Good	Excellent	Excellent	Excellent	Fair

Figure 9.15 Comparison of various types of conventional pressure sensors for internal and external measurements.

As other technologies develop, the suitability of other types of sensors will improve. Optoelectronic sensors will likely become smaller and less expensive in the future. Alternative packaging techniques for IC pressure sensors will allow them to be used to measure pressure between compliant surfaces, and signal conditioning circuitry can be built-in. IC sensors have the further advantage that multiple sensing technologies, such as temperature and pressure, can be included on a single chip. Development costs are significant, but batch fabrication can reduce parts cost.

9.9 STUDY QUESTIONS

9.1 Explain the difference between a pressure sensor and a force sensor.

9.2 Explain the construction of conductive polymer pressure sensors. Sketch the resistance vs. pressure curve.

9.3 For conductive polymer pressure sensors, explain the difference between shunt and through-conduction modes. Explain methods of attachment.

9.4 Explain the cause of the piezoresistive effect.

9.5 Design a complete circuit for a one-active element strain gage and amplifier.

9.6 Design a complete circuit for a capacitive pressure sensor.

9.7 Which variable in the capacitance equation is generally changed in capacitive pressure sensors?

9.8 List advantages and disadvantages of optoelectronic sensors.

9.9 Explain the difference between intrinsic and extrinsic fiberoptic sensors. Which are more structurally reliable?

9.10 What are advantages of measuring interstitial pressure by the wick catheter method over the needle manometer method?

9.11 List important design considerations for implantable sensors.

9.12 Compare the advantages and disadvantages of various types of conventional pressure sensors for interface pressure measurement.

10

Interface Pressure Distribution Visualization

Annie Foong

Chapter 9 discussed different types of sensors available for pressure measurements. This chapter will examine the techniques available in providing an overall visualization of pressure distribution—explicitly or otherwise. Both qualitative and quantitative techniques will be reviewed.

10.1 VISUALIZATION BY DIMENSIONAL CHANGES

10.1.1 Qualitative Methods

One of the simplest ways to qualitatively visualize pressure, by dimensional changes, was by accessing indentations made in mud, clay, and other similar materials (Betts and Duckworth 1978). However, such methods tend to measure the shape of the loading object instead of pressures. Hughes *et al* (1987) reported a technique used by Harris (1947), which produced an ink print of pressure patterns. A rubber mat with three layers of fine ridges was covered by a layer of ink. By applying pressure on the mat, a corresponding imprint pattern is produced on a piece of paper placed below the mat. However, this method did not have very good pressure resolution and basically produces imprints only for three precalibrated pressures.

10.1.2 Dimensional changes in supporting surfaces

Bed of nails and springs
Figure 10.1 shows a "bed of springs and nails", proposed by Lindan (1961), for measuring pressure distributions over the surface of a supine human body. He made use of the principle of spring compression, relating the deformation of springs to the load applied. The supporting surface consisted of "nails" mounted on multiple independent calibrated springs whose compression could be measured. The bed contained up to 1000 such nails and springs. The nails/springs were placed about 1.4 to 2.0 cm apart except at sharp prominences at the heels, where they were placed closer together. The force applied on each nail was obtained using Hooke's Law ($F = kx$ where F = force, k = spring

142

constant, x = spring displacement). The pressure was then obtained by dividing the force by the area of interest.

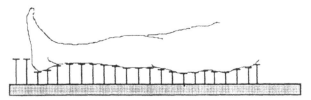

Figure 10.1 A supine subject on the "Bed of Springs and Nails" permits visualization of spring compressions. Springs are not shown.

Example 10.1
Given that the spring constant is 160 N/m, the nails are situated 2 cm apart, and are compressed 1 cm, what is the pressure at that point ?

Force = $F = kx$ = (160 N/m)(0.01 m) = 1.6 N
Pressure = Force/Area = 1.6 N/4 cm^2 = 4 kN/m^2 = 4 kPa = 30 mm Hg.

The hardness of the surface can be varied by using different spring constants or by varying the nail spacings. Hence, by careful variations of these parameters, we can obtain surfaces of required compliance. The bed that Lindan used had a surface that corresponded roughly to hard surfaces padded with 75 mm thick foam rubber commonly used in hospitals.

Operating such a device can be a tedious process. Measurements must be taken manually for all 1000 nails and subjects are required to remain on the bed 1 to 2 h in order for the measurements to be completed. As such, this device cannot be used on patients already prone to pressure sores. There is also a problem of individual nails being wedged (by lateral forces), and this must be corrected by first pulling down on individual nails and then releasing. Moreover, nails that are not at the peripheral of the body are difficult to measure. Lindan claimed that having taken these precautions, 90% of the body weight rests on the nails. What is worth noting is that this device was able to measure pressure distribution of a supine subject. The nails faithfully hug the body contour and gave a pressure reading at each point. Variations of this method have appeared in Cooper *et al* (1986) where Hall transducers were used to automatically detect the nail displacement.

Potentiometers
Figure 10.2 shows a contour gage which had a sliding potentiometer for displacement measurements (Reger *et al* 1985). The sliding potentiometer was fastened to an aluminum tube probe which extended through predrilled 9 mm diameter holes. The top of the tube was attached to the cushion cover and any vertical displacement was transmitted to this probe causing a change in output voltage. By assembling a matrix consisting of 64 (8 × 8) such probes, a three-dimensional support surface over the entire matrix can be obtained. Outputs from the potentiometers are multiplexed from 64 to 4 channels into 12-bit ADCs, with a sampling rate of 3.3 MHz, into a data acquisition system.

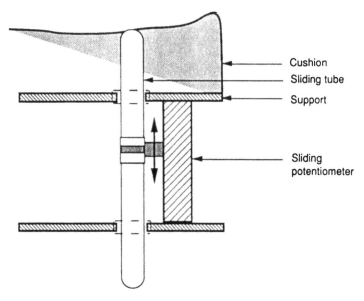

Figure 10.2 The sliding potentiometer yields a output voltage indicating contour displacement (Reger *et al* 1985).

The entire system was interfaced with the LSI-11/23 computer, for automatic data acquisition. A 6th-order polynomial curve-fitting algorithm was used to interpolate between the discrete data points. Reger claimed that there was no major change in the support characteristics of the cushion, as only about 2.5% of the cushion material was removed to house the sliding probes.

In addition, a pressure pad, very similar to the TIPE pad discussed in Section 8.3.3, was also interfaced to the computer, and used to simultaneously obtain pressure readings. Hence, it was possible to directly map the interface contour deformation to its corresponding pressure measurements. Software, capable of data acquisition, analysis and display, was developed to accommodate the added complexities.

Ultrasound

Figure 5.9 shows how ultrasound was used to quantitatively measure cushion indentation contour at the buttocks–cushion interface. The ultrasonic contouring system (Kadaba *et al* 1984, Laliberte 1985) was based on the fact that an ultrasonic pulse traveling in a multilayered medium is reflected and refracted at each interface between two layers of different acoustic impedances. By noting the time delay between a transmitted pulse and its returning "echo", and knowing the speed of ultrasound in that medium, the investigators were able to obtain dimensional changes in a loaded cushion. Kadaba obtained "slices" of pressure distribution (figure 10.3) by manually moving an ultrasonic transducer along a slot in 5 mm steps, and concurrently recording the returning echoes.

Laliberte (1985) reported an improved version of the ultrasonic contouring system. The positioning of the ultrasound transducer was accomplished through the use of two precision sliding assemblies driven by a stepper motor, and controlled by an IBM CS/9000 computer. As a result, the transducer can be positioned very accurately at any required (x, y) coordinate. The contour display was obtained as follows: as the transducer moved along a scan line, in the x-direction, each new contour point was displayed until the transducer reached the end of that line, thus giving a contour slice at any given y (figure 10.3). The transducer was then advanced to a new y position, and a new scan was again performed. At completion, a three-dimensional contour plot was then obtained by displaying the various contour slices simultaneously on the display screen (figure 10.4).

10.1.3 Dimensional changes in tissues

Reger *et al* (1986) used the technique of Magnetic Resonance Imaging (MRI) to obtain cross-sectional views of the ischial tuberosities during loading. MRI

Figure 10.3 A slice of an ultrasonic scan displays buttock–cushion contour.

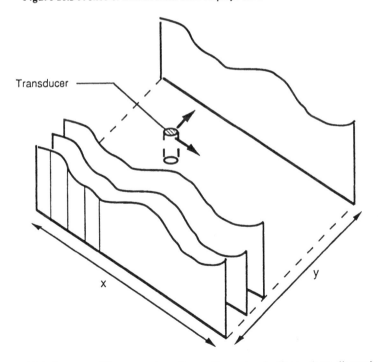

Figure 10.4 By assembling a series of two-dimensional slices, three-dimensional visualization of the buttock-cushion contour is possible.

allows images of the cross-sectional anatomy to be visualized conveniently and noninvasively. Tissue deformation was quantitated in the 0 to 60 mm Hg pressure load range with approximately 1 mm^2 spatial resolution (Reger *et al* 1986). Subjects lie supine in the magnetic field of the imaging system supported by 65 mm thick polyurethane foam cushions. External pressure was measured under the bony prominences before and after each imaging run using the Scimedic pressure system. The corresponding loading force was calculated from the measured pressure and the area of the Scimedic transducer bladder (65 cm^2). By visualizing MRI images on a CRT, tissue thickness, from bony prominence to the skin surface, can be obtained using cursor controls. Thickness measurements were taken of both loaded and free-hanging tissues, from which calculations are made of tissue stiffness:

Tissue deformation= Thickness (free-hanging) − Thickness (compressed) mm.
Tissue stiffness   = Load/tissue deformation N/mm.

Reger found that the paraplegic subject has muscle stiffness of 3.4 N/mm and a skin stiffness of 6.8 N/mm. The normal subject has muscle stiffness of 1.4 N/mm, and an interestingly anomalous skin stiffness of −8.7 N/mm. In other words, normal skin tends to increase in thickness under loading. No explanation was given for this anomaly. A plausible reason could be that normal skin is very viable and compliant, and the ability to increase in thickness under load is the body's natural response to protecting loaded areas. More importantly, he found that foam cushions, in the 0–25% compression range, have a stiffness of about 3.6 N/mm, which is a reasonable match for muscles but too low for skin. For cushion compressions larger than 25%, however, cushion stiffness increased considerably, making the discrepancy smaller at the skin but larger for muscles. Therefore, by using MRI visualization techniques, Reger concluded that it is important to note the precise matching of the shape and material of support surface to the buttocks, in order to preserve as many of the seating interface characteristics as possible.

10.2 VISUALIZATION BY AN ISCHIOBAROGRAPH

The original barograph was developed by Chodera in 1957 and several adaptations of that device have since appeared. The pedobarograph is widely used by investigators for measuring pressures under the foot (Betts and Duckworth 1978, Rhodes *et al* 1988, Hughes *et al* 1987). When used for visualization of pressure measurements under the ischium, the device is aptly known as an ischiobarograph.

10.2.1 Principle of operation

The basic system (figure 10.5) consisted of a glass plate, usually crown glass, illuminated at both ends by fluorescent lamps. The top surface was covered with a thin sheet of opaque reflective plastic, usually 100 μm white PVC. The plastic should be opaque to eliminate light coming from above. Light from the fluorescent lamps emerging from the glass plate increased with the amount of pressure applied to the plastic. These changes can be recorded by a camera or any appropriate imaging system placed beneath the device.

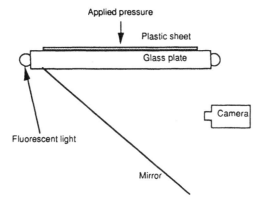

Figure 10.5 An ischiobarograph utilizes a plastic sheet on an illuminated glass plate (Betts and Duckworth 1978).

Figure 10.6 shows the principle of operation of the ischiobarograph. Light from the fluorescent lamps enters at angles, which causes total internal reflection between glass–air surfaces. For total internal reflection to occur, the light ray must be traveling from a medium with high refractive index to one with a low refractive index, and hit the surface at an angle greater than the critical angle for the materials concerned.

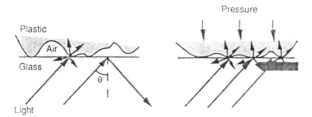

Figure 10.6 Light entering the barograph is totally internally reflected when θ, the angle of incidence, is greater than θ_C, the critical angle, which is 42° for crown glass. μ is the refractive index, $\mu_{air} < \mu_{glass} < \mu_{plastic}$ (Betts and Duckworth 1978).

On a microscopic scale, the plastic has an uneven but deformable surface. Some investigators (Minns *et al* 1984) used a plastic with serrated indentations. Under unloaded conditions, the presence of a large number of air gaps between the plastic and glass causes most of the light to undergo total internal reflection within the glass. When pressure is applied, the plastic sheet comes in closer contact with the glass—the greater the pressure, the more intimate the contact. As a result, the internal reflection of light is broken and more light emerges from the glass plate. The light intensity pattern, therefore, provides an indication of pressure distribution.

To obtain useful quantitative data on pressures, video information from the camera is converted into pressure contour maps through the use of a color interface system, consisting of comparators. The comparators split the video signal into various levels, and assign a pseudo color corresponding to a particular pressure value. The resultant display is a color pattern, giving the pressure distribution as a contour map

10.2.2 Related implementations

Pressure profiles for other surfaces

The most obvious shortcoming of the ischiobarograph is that it produces pressure profiles only for hard surfaces—thin plastic on a hard glass plate. However, Mayo-Smith (1980) did report a modified wheelchair version of the ischiobarograph where a rubber mat was used. The rubber mat had a smooth top surface, but with equally spaced inverted pyramids on the bottom. The pyramids' apexes contacted the glass plate and served the same purpose as the plastic sheet in conventional ischiobarographs. When pressure was applied on the rubber mat (figure 10.7), the pyramids' apexes flattened against the glass plate, broke down the internal reflection in these regions, and produced spots of bright light. In other words, the apexes of the rubber mat served as the serrated indentations of a conventional ischiobarograph plastic. The indentations, in this case, were at a macro, rather than a micro level. The investigators' main concern was to locate regions of high pressures and they did not report any quantitative measurements. However, it is easily conceivable that similar pressure contour maps for the modified device could also be obtained. However, resolution may not be as good as that with a conventional ischiobarograph because larger areas of contact were produced by the pyramids' apexes, giving average pressure readings over the area.

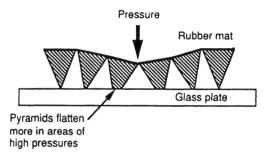

Figure 10.7 Loaded rubber mat with flattened pyramids in areas of high pressures. Internal reflection of light is broken down in these areas, producing bright spots of light (imaging setup similar to that of figure 10.5 is not shown here).

Use of a photoelastic plastic

Rhodes *et al* (1988), in their investigation of ground–foot reaction forces, reported a device whose setup is very similar to the ischiobarograph. Pressure is similarly visualized as patterns of light intensity emerging from the device. However, instead of using a simple plastic–glass interface, they made use of a piece of photoelastic material (polyurethane) to form their force plate.

A photoelastic material has the property of becoming birefringent when loaded. Birefringence in a material causes polarized light to be resolved into two orthogonal directions. Because the velocities of light propagation are different in each direction (McGraw-Hill 1987), the two components emerge out of phase from the photoelastic material. The relative retardation of these components is directly proportional to the stress applied. Such a phenomenon is usually visualized through a device known as the polariscope (figure 10.8). Light is first polarized by a polarizer, travels through the photoelastic material where it is resolved into the two components, and is brought together again by a second polarizing element (analyzer). At the analyzer, an interference pattern of bright and dark fringes emerges. Unstrained areas appear dark, and strained areas appear bright (Rhodes *et al* 1988). If white light is used, instead of monochromatic light, the relative retardation causes fringes to present themselves as colors of the spectrum—the photoelastic material acting as a sort of prism. White light is often used for demonstration purposes, monochromatic light for more precise measurements (McGraw-Hill 1987). The transmission-type polariscope (figure 10.8) has the polarizer and analyzer on opposite sides of the photoelastic material. A reflective polariscope, similar to the one used by Rhodes *et al* (figure 10.9), has only one polarizing element to serve as both the polarizer and analyzer. Reflection is effected by coating the photoelastic material with a mirrored surface.

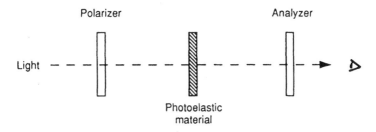

Figure 10.8 A transmission-type polariscope.

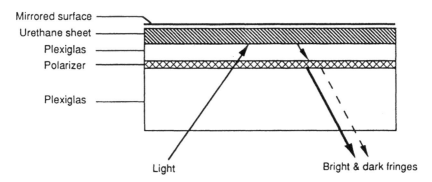

Figure 10.9 A reflective-type polariscope, similar to the force-plate used by Rhodes *et al* (1988) (imaging apparatus not shown here).

In the device of Rhodes *et al*, the top surface of the force plate, which replaced the plastic–glass interface of a conventional ischiobarograph, is painted silver to reflect light illuminating the sheet from below. A piece of urethane sheet served as the photoelastic material. The resulting fringe pattern was captured by a 256-intensity level imaging system, interfaced to an IBM AT. These intensities were displayed as pseudo colors for better visualization. As with a conventional ischiobarograph, precalibration was necessary to relate pressures to light intensities.

Other literature has shown that with more involved techniques and added calculations, it is possible to use photoelasticity for visualizing stresses in three spatial directions—shear and compressional stresses (McGraw-Hill 1987). We have found no studies where investigators made use of such a principle to measure shear forces. However, with more emphasis being placed on the importance of shear forces as contributory to pressure sores, it is possible that a variant of the device of Rhodes *et al* may be used to visualize shear forces. Other photoelastic materials include Bakelite, celluloid, gelatin, and certain synthetic resins (McGraw-Hill 1987).

Other problems

The relationship between pressure and intensity is nonlinear and precalibration is necessary. Since the calibration procedure requires a uniform pressure loading system, an air-pressure system is used. The calibration process can be extremely tedious (Rhodes *et al* 1988). Without calibration, the ischiobarograph is good for making quick qualitative visualization of pressure distributions. The visualization process is further aided by the pseudo color patterns presented.

In addition, the plastic used must have good dynamic range. It must also be able to recover quickly when pressure is released and not stick to the glass surface. Most plastics which have good responses to pressure tend to stick to the glass when pressure is removed and leave a greasy deposit (Betts and Duckworth 1978). Conversely, plastics which do not stick to glass have poor responses to pressures. A compromise must therefore be reached. Minns *et al* (1984) reported that for normal human loads, most materials settle to an equilibrium within 2 min.

It is also important to consider the absorption properties of glass. Illumination from only one glass edge results in fall-off errors, i.e. areas nearer to the light source receive more light than areas further away. Therefore, two lamps are used, each at one side of the glass edge. Two-lamp illumination, yields a central area where illumination is uniform. Pressure measurements should only be taken in this region of uniform illumination to avoid fall-off errors. Finally, present ischiobarographs are unable to measure shear forces.

10.3 VISUALIZATION WITH COMPUTER DISPLAYS

Adding a computer to a system can greatly enhance its capabilities. Data acquisition and storage is made possible, thereby providing a better ability to process data. Furthermore, supplemented with powerful software, more sophisticated visualization is possible. It is conceivable that computing power, and a corresponding interface, can be added to any of the systems described previously. In the ischiobarograph, for example, the patterns of light intensities can be captured via a camera linked to an imaging card. Pixel intensity conversions to pseudo colors, automatic edge detection of isobar lines, noise filtering, etc., can be achieved using common image processing techniques.

A well established computerized system is that of the SPIRAL pressure monitor (Jaros 1987). The SPIRAL pressure monitor consisted of three parts: the TIPE sensor pad (section 8.3.3), an electronic interface, and software. The interface was a proprietary circuit board meant to be plugged into the IBM PC. This board had one circuit for monitoring the state of individual pad switches and a second circuit to read the internal air pressure of the pad. The simplest level of visualization was a matrix of LED displays—each LED corresponding to the ON/OFF status of a cell on the TIPE. The SPIRAL software provided improved visualization. It continuously monitored the state of the hardware switches and updated the computer's video display grid. A matrix of graphical squares were then displayed—glowing red to indicate a closed switch and faint blue otherwise. Such a graphic display is useful in providing the user with a fast visual feedback. Finally, the program was also able to provide a two-dimensional seating pressure distribution map which gave the pressure distribution over a short period of time (figure 10.10). This was accomplished by first inflating the pad until all switches were opened and then recording as the pressure was slowly lowered. The software monitored when the switches reached their "first-closed" state and recorded the corresponding pressure. Note that the pressures shown in the distribution map of figure 10.10 do not change "smoothly" from one cell to another. This is probably due to the fact that the readings are not done at a single instant, but rather over a finite period of time, during which the pad was deflated slowly. As such, a shift of weight might have occurred and altered the pressure distribution. This explanation is further supported by the fact that the software actually has the capability to monitor pressure distribution over several recordings to provide an average map so as to minimize errors.

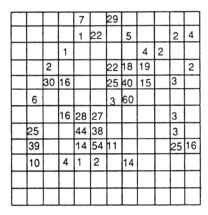

Figure 10.10 The TIPE pad and SPIRAL monitor provide a two-dimensional pressure distribution map with values in mm Hg (Jaros *et al* 1988).

Babbs *et al* (1990) constructed a mat that can sense pressure in a 24×64 array of capacitance pressure sensors. Figure 9.9 shows the principle and figure 10.11 shows the gray scale image of the pressure distribution.

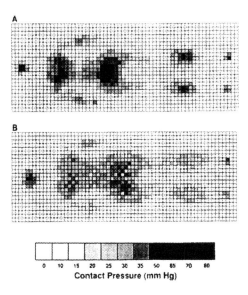

Figure 10.11 (a) Image of pressure distribution created by a supine adult female subject on a hard wooden surface. The pressure-sensitive mat was interposed between the subject and a wooden table. High-pressure areas associated with the head, shoulder blades, sacrum, calves, and heels are evident (left to right). (b) The same subject on a soft surface. The pressure-sensitive mat was interposed between the subject and a 4-inch-thick foam pad. Attenuation of high-pressure points is evident, together with spreading of pressure over a larger area (Babbs *et al* 1990)

In our investigations of the topic, we used a 4×4 Interlink sensor array for measurements of pressure distribution under the ischial tuberosities on a hard surface. The array was interfaced to a data acquisition board (Tekmar board) on an IBM PC. Having acquired the necessary data, these were fed into a graphics program (Surfer), capable of curve fitting and data interpolation, to produce the three-dimensional plots shown in figures 10.12 and 10.13. The plots show pressure distribution over the left tuberosity.

We have also constructed an 8×8 array of discrete Interlink sensors sandwiched between layers as suggested by Webster (1989). The base was a $0.025 \times 12 \times 12''$ pure gum mat. Sensors were glued on top of 0.025'' aluminum metal backups to minimize hysteresis. Transparent plastic formed the top layer. Figure 10.14 shows the resulting real-time computer-generated three-dimensional plot generated by an IBM PC.

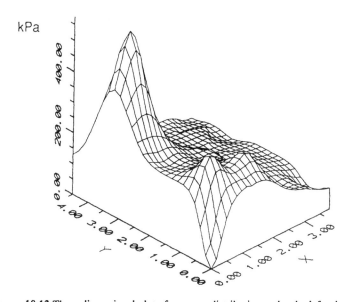

Figure 10.12 Three-dimensional plot of pressure distribution under the left tuberosity. Data point at ($x = 0$, $y = 0$) had been forced to 0 kPa, in other to reveal subtle changes in the pressure profile of nearby regions (x and y axes are given in cm).

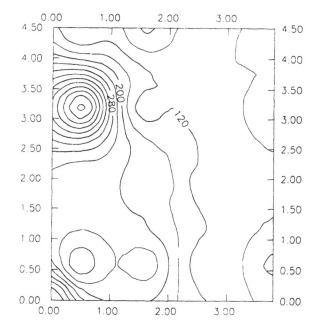

Figure 10.13 Contour map with similar data from figure 10.11.

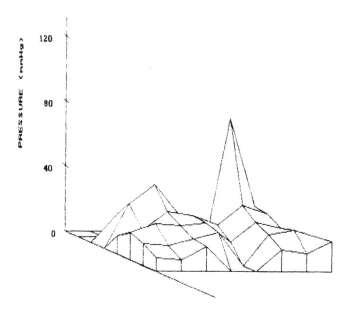

Figure 10.14 Three-dimensional plot of pressure distribution derived from an 8 × 8 array of Interlink pressure sensors placed between the buttocks and the top of an office chair containing a 1-cm foam pad.

10.4 OTHER VISUALIZATION TECHNIQUES

Visualization can also be achieved through the use of thermograms. High-pressure areas have higher temperatures than lower pressure areas. The reader is referred to Chapter 3 for further details of thermography and its relationship to pressure distributions.

10.5 STUDY QUESTIONS

10.1 Describe the bed of nails method of measuring contour and how to present its results.
10.2 Suggest an alternative method of measuring support surface displacement that is not described in this chapter.
10.3 Describe the advantages and disadvantages of using an ischiobarograph for measuring pressure distribution.
10.4 Explain the principle of total internal reflection as used in the ischiobarograph.
10.5 Give a block diagram design of the comparator circuit interface described in section 10.2.1.
10.6 Give a plausible program design summarizing the SPIRAL software.
10.7 Explain the inverse-weighted-distance algorithm.

11

Accuracy of Pressure and Shear Measurement

David Beebe

The accuracy of pressure and shear measurements is an often overlooked subject. In clinical settings, pressure measurements are obtained by means of a wide variety of sensors. Unfortunately these sensors have different characteristics, giving rise to a variety of errors. Measurement procedures also vary between institutions. In this chapter we will examine the factors that can affect the accuracy of interface pressure measurements. Calibration methods and techniques will also be discussed. Finally, the importance of shear forces in the formation of pressure sores will be discussed and a variety of methods for measuring shear will be examined.

11.1 PRESSURE MEASUREMENT

11.1.1 Sensor calibration

In order to make use of any sensor to measure some quantity, it is first necessary to calibrate the sensor. But what does it mean to calibrate a sensor? The dictionary definition of calibrate is to determine, check, or rectify the graduation of any instrument giving quantitative measurements. Applying this definition to the calibration of a pressure sensor gives us the following working definition: obtain sensor input and output data over a range of values, then use statistical methods to perform curve fitting on this data resulting in a calibration curve (Jensen 1990). The goal of calibration is to produce a plot of sensor output vs pressure. This plot or calibration curve is then used to convert sensor output to pressure values when measuring unknown pressures.

Calibration techniques
We can calibrate sensors using one of several different methods, chosen based on the intended use of the sensor. In other words, we should attempt to calibrate the sensor under conditions identical to the conditions present during the intended use of the sensor. First, we must decide how to collect the sensor output data. The two types of data collection are commonly known as static and

dynamic data collection. If the data are taken statically, data are recorded only when the sensor output is fixed and stable. If the data are taken dynamically, data are recorded (usually at fixed time intervals) while the sensor output is changing. Since we are mainly interested in measuring static pressure distributions at the interface between the patient and the support material, static data collection should be used.

Next, we must decide whether to carry out the calibration using manual or automatic techniques. The following is an example of a manual calibration of a simple conductive polymer pressure sensor. First, apply a known pressure to the sensor and use an ohmmeter to measure the resistance. Next, repeat this operation over the range of pressures expected during the intended use of the sensor. Finally, manually plot the data and use regression analysis or interpolation to fit a curve to the points. Manual calibration has the following disadvantages: (1) very slow, (2) limited to static data collection unless output is changing very slowly, (3) subject to accuracy and consistency of experimenter. The principal advantage of manual calibration is its simplicity; no special equipment is required.

The following is an example of a automatic calibration of a simple conductive polymer pressure sensor. First, apply a varying pressure to the sensor while using a computer and analog-to-digital converter (ADC) to simultaneously collect and record the sensor output. In this case the varying force is simultaneously applied to the sensor to be calibrated as well as a high-accuracy reference sensor. Next, use computer regression algorithms, such as Minitab, to fit a curve to the points. Automatic calibration has the following advantages: (1) it is fast, (2) it works for both static and dynamic data collection, (3) higher order regression analysis is possible. A disadvantage of automatic calibration is the increased complexity of the required equipment. However, the advantages of an automatic calibration system far outweigh the disadvantages.

Regression analysis or the method of least-squares is the most common technique used in fitting a curve to a set of data points (Doebelin 1990). The principle behind the least-squares method is to minimize the sum of the squares of the vertical deviations of the data points from the fitted curve. A variety of commercially available software exists to carry out regression analysis. The form of the equations found using regression is

$$P = b_0 + b_1 S + b_2 S^2 + b_3 S^3 + \ldots$$

where

P = Pressure applied to the sensor,
S = Sensor output,
b_n = Calibration coefficients from the regression.

Ideally, sensor calibration should be carried out under conditions identical to those under which the sensor will ultimately be used. For example, if a sensor is calibrated on a flat surface, but in its intended use it will be mounted on a curved surface, errors may result. Environmental conditions, such as temperature and humidity, should be considered.

Hysteresis

Hysteresis results when some of the energy applied to the sensor for an increasing input is not recovered when the input decreases (Olsen 1978). Figure 11.1 shows a typical calibration curve of a sensor that exhibits hysteresis. That is the output at particular pressure is different during loading and unloading. The output tends to stick at the old value and lag behind where it would be for a sensor without hysteresis. Hysteresis is often dependent on the rate of change of the input. This is another reason why the sensor should be calibrated under conditions identical to those under which the sensor will ultimately be used.

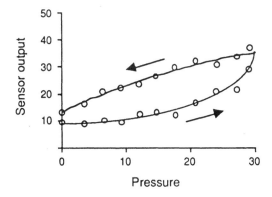

Figure 11.1 A typical calibration curve of sensor output vs actual pressure. The o's are sensor output data, the solid line is fitted to the data points using regression analysis, the arrows indicate the direction of the hysteresis. Arbitrary units are used.

Calibrating pressure sensors

Each different type of pressure sensor has its own unique characteristics. As a result, the calibration methods used will vary depending on the sensor of choice. When calibrating an conductive polymer pressure sensor, a simple brass weight can be used to apply a uniform pressure to the sensor. Since the sensor area is known the applied pressure is easily calculated. However, applying a simple brass weight to a semiconductor diaphragm pressure sensor will produce no change in the sensor output. Figure 9.3 shows that placing a rigid weight with an area larger than the diaphragm will not cause any deflection of the diaphragm. Figure 11.2 shows a simple calibration apparatus which has proved useful for calibrating a variety of pressure sensors. A pneumatic pressure chamber driving a piston is another common calibration apparatus.

Calibrating shear sensors

Calibrating shear sensors can be accomplished using a set up shown in figure 11.3. The units of horizontal shear are pascals (N/m^2), where the force is applied horizontally, and the area is the area over which the shear is applied to the sensor. By comparing the area of weight A with that of the shear-sensing element in contact with the simulated tissue, the area of the applied shear can be found. This test set up can also be used to evaluate the interaction of pressure and shear, and the motion error of the sensor as discussed in Section 11.3.2.

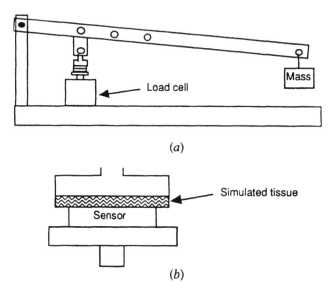

(a)

(b)

Figure 11.2 (a) A typical calibration system consisting of a lever arm and attached weight, load cell, and ball-joint-mounted load applicator. (b) Expanded view of sensor, applied load and support. Note simulated tissue is used in an attempt to replicate the intended use of the sensor.

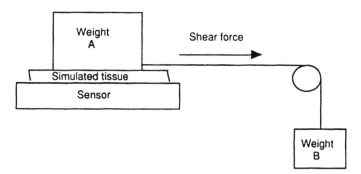

Figure 11.3 A system for calibrating shear sensors. Weight A is the applied pressure, weight B is the applied shear. The area of contact between the simulated tissue and the sensing element is the area of the applied shear (Bennett *et al* 1979).

11.1.2 Sensor accuracy

Sensor accuracy can be affected be a variety of external influences, such as temperature, time, humidity, and electromagnetic interference. For example, semiconductor strain gages have a strong temperature dependency. Because of this strong temperature dependency, semiconductor strain gages are usually used in a Wheatstone bridge configuration so that the temperature effect tends to cancel. Bennett (1990) encountered drift caused by skin moisture with some pressure sensors. When using any sensor we should carefully examine all its properties and modify its use so as to minimize or compensate for the effect of external influences.

11.1.3 Sensor cost

Many different sensors have been used to measure pressure distributions. Figure 11.4 shows that the cost of these sensors vary from a few dollars for a simple conductive polymer sensor to thousands of dollars for unique experimental sensing systems.

Sensor	Cost	Advantages	Disadvantages
Bladder array	?	Simple, good repeatability	Cushion effect
Conductive polymer	$1	Thin, overforce protection, cost	Hysteresis, creep
Semiconductor strain gage	$10–600	High sensitivity, repeatable	Available packaging, temperature dependent
Metal strain gage	$5–200	Linear	Low sensitivity
Capacitive	$10–200	Small size	Drift, hysteresis
Optoelectronic	$10–2000	Noise immunity, easily multiplexed	Fragility of fibers, thick
Implantable wick catheter	?	Measure internal tissue pressure	Must be inserted into the tissue

Figure 11.4 Sensor costs vary widely and are not necessarily a function of performance. Refer to Chapters 8 and 9 for a complete description of the sensors listed.

11.2 PRESSURE MEASUREMENT INCONSISTENCIES

11.2.1 Factors affecting results

Often when interface pressures are measured, a pressure sensor is placed between the load and the support, and its output is assumed accurate and reliable. Unfortunately, this assumption is often wrong. The accuracy of the pressure measurement can be affected by a variety of factors. McGovern *et al* (1987) lists the following major factors affecting interface pressure evaluations: (1) the sensor size and material, (2) the load shape and its interaction with the support material, (3) the method of endpoint detection, and (4) the uniformity of the measurement technique.

Sensor size and material
All sensors have a finite thickness which creates a gap between the load and the support surfaces (Reddy *et al* 1984, Vinckx *et al* 1990). Due to this "gap" or perturbation of the normal load–support interface, the sensor tends to support the load. This causes local concentrations of stress, with the intensity of the concentrated stress depending on the compliance and thickness of the sensor.

Two approaches are possible when attempting to minimize the effect of sensor size on the results of interface pressure measurements. The perturbation of the load–support interface tends to be minimized if the sensor area is either very small, or is equal to the total area of the interfacing surfaces. In other words, the sensor should either support all of the load or should share it uniformly with the support surface, in effect matching its mechanical impedance. A force sensor results if a single sensor supports the entire load and the area of the applied load is equal to or larger than the sensor. So to measure pressure distributions, a small sensor, relative to the load size, should be used. Regardless of the sensor size, its material properties will also affect the measurement. Ideally, the sensor should have an infinitesimal thickness and its structural properties should exactly match those of the load or support. A more realistic design which would minimize the errors due to sensor size and material would be an array of small sensors which are an integral part of a thin elastic pad which could be placed between the load and the support. Unfortunately, no existing sensors meet these stringent requirements. As a result, if absolute pressures are required, a correction factor will be necessary when using existing pressure sensors.

Load shape and its interaction with support material
Patterson and Fisher (1979) examined the performance of five pressure sensors under three different loads. Initially the sensors were evaluated using rubber and cloth surfaces for loading. The sensors were placed on the inside surface of a tube 3 cm in diameter with the sensor's active surface facing inwards. A condom was placed in the tube and inflated in 20 mm Hg increments up to 200 mm Hg. Simultaneous measurements were taken of the air pressure in the condom and the pressure indicated by the sensors. The same procedure was repeated with a soft cotton cloth 0.3 mm thick placed in the cylinder between the condom and sensor. Figure 11.5 shows the results. Good agreement results when using the condom only. However, when the cotton cloth is added, large errors result. In fact, the output of one sensor actually goes negative. The

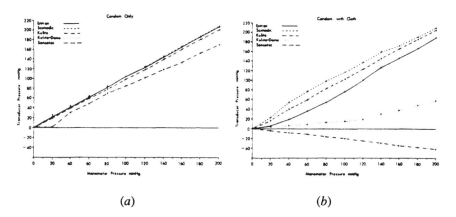

(a)   (b)

Figure 11.5 (a) There is good agreement between condom and sensor pressure measurements. (b) The introduction of a cloth interface produces large errors (Patterson and Fisher 1979).

precise cause of theses errors is not known, but Patterson (1990) suggests that the cause is related to the uneven loading of the sensor diaphragm by the cloth. The sensors are designed for use with fluids where the load on the diaphragm is uniform.

Next the sensors were placed in a line across the width of a 13 cm wide blood pressure cuff. The sensors were placed 1.5 cm apart and 3.5 cm from either edge. The blood pressure cuff was then wrapped around the calf of the leg. Figure 11.6 shows the four test locations around the calf. The results show that in general the errors increased as the sensors were moved from position 1 through position 4. The subjective feeling is that the underlying tissue becomes harder or more "bony" from the gastrocnemius muscle position to the tibial crest position. A particularly large variance of data was obtained over the tibial crest. This is attributed to two factors: (1) the inability to exactly and consistently place the sensor precisely on the crest of the tibia, (2) unequal loading of the sensor due to the irregularity of the crest surface.

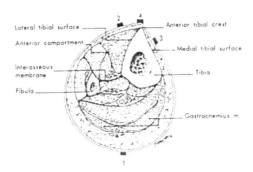

Figure 11.6 Cross section of the calf showing the location of the pressure sensors. Measurements were made from position 1 to 4 reflecting an increasingly firmer underlying tissue (Patterson and Fisher 1979).

Endpoint detection
Defining the useable range or endpoints of a sensor is very important. Often a sensor will continue to provide an output well beyond its useable range. For example, a simple conductive polymer pressure sensor will tend to saturate under excessive pressures (i.e. the output will stop changing at high pressures). At low pressures the conductive polymer output is also limited. A spacer exists between the conductive layers, because of the space there will be no output until some minimal pressure is applied. So we must be careful to operate within the range of operation of the sensor or the results may be skewed.

Measurement technique
Measurement techniques are developed independently at a variety of institutions. This wide variety of techniques and procedures gives rise to a wide variety of results which are often hard to compare. In addition sometimes the procedures used are poorly designed, thus giving rise to experimental errors. In section 11.2.2, a study by Bader and Hawken (1986) is used to illustrate some of the problems produced if measurement techniques and procedures are not

carefully considered. Then in section 11.2.3, proper experimental design will be discussed.

11.2.2 Effect of reseating

Bader and Hawken (1986) showed large variations in pressure measurements when the subject was reseated between pressure readings.

Instrumentation
Measurements were taken using a matrix of twelve bladder type sensors 20 mm in diameter, spaced 28 mm between centers. In order to maintain the sensor matrix in a fixed anatomical position, the matrix was centered over the subject's right ischial tuberosity and was attached at the edges to the skin with medical adhesive tape. The matrix was attached with the subject in a semikneeling position with both knees and hips flexed. This posture is similar to the seated posture, and ensures that the matrix remains in contact with the interface throughout the test without wrinkling or overstretching. The subject was seated in an specially designed assessment chair. The chair had an adjustable seat height, hip pads, seat depth, and back support profile.

Procedure
Each subject was seated on a foam cushion and asked to remain as still as possible while five sets of measurements were taken. It took approximately 60 s to record each set of measurements. The subject then stood up for a sufficient time to allow the cushion to recover its initial thickness. Finally the subject was carefully reseated in the same position. This procedure was repeated five times.

Results
Pressure measurements taken during one seating showed small variations (a few mm Hg over intervals of several minutes). However when comparing the measurements taken at different seatings, large variations were found. The Kruskal-Wallis analysis (Agresti and Agresti 1979) was used to test the hypothesis that the variability in pressure produced by reseating the subject was no greater than the variability occurring while the subject remained quietly seated. The analysis showed that in 19 out of 20 cases the hypothesis was rejected at the 5% level at least. In fact the rejection level was 0.1% in 9 of the 20 cases. A rejection level of 5% means that there is a less than 5% chance that the pressure readings between seatings will fall in the same range as those readings obtained while the subject was sitting still. In other words, the variability in pressure produced by reseating the subject was significantly higher than the variability occurring while the subject was sitting still. This is a very important result for two reasons. First, the experimenters exercised great care when reseating the subjects. A deliberate attempt was made to reseat the subject in exactly the same position each time. The results emphasize the fact that the subject can take up any one of a large number of equilibrium postures, each of which involves a different balance of muscular and supporting forces. Even though the differences are not readily perceived by the experimenter or subject, the small changes in supporting forces were reflected in the significant variations in pressure distribution. Second, if tests of several different support

surfaces are performed, these repositioning changes may swamp the differences due to the support surface. Proper experimental design can help alleviate this problem. Several sets of pressure readings should be taken for each support surface and between each set of readings the subject should stand up and then sit down again. Thus variability due to repositioning will exist both within and between groups and normal statistical methods will detect any extra variability that can be attributed to the different surfaces.

11.2.3 Experimental design

Engineers often concentrate on the technical aspects of an experiment while overlooking experimental protocol. If an experiment is poorly designed, the results may be misleading and the amount of useful information gained minimal. There is an enormous amount of information available in the field of experiment design and statistical analysis of experiments. The intent here is not to make the reader an expert on the subject, but rather to convince the reader of the importance of proper experimental design and analysis.

Replication and randomization
Replication and randomization are the two fundamental principles of experimental design (Montgomery 1976). Replication has two important properties. First, it allows the experimenter to obtain an estimate of the experimental error. This estimate is important in determining whether observed differences in the data are really statistically different or due to experimental error. Second, a sample mean or average is normally used to estimate the effect of a factor in the experiment. Replication allows the experimenter to obtain a more precise estimate of the effect. What do we mean by randomization? Randomization means that both the allocation of the experimental material and the order in which the individual runs or trials of the experiment are to be performed are randomly determined. Proper randomization will tend to minimize or average out the effect of extraneous factors that may be present.

Single factor vs factorial experiments
Experiments can be divided between single factor experiments and factorial experiments (Box *et al* 1978). Whenever only one factor is varied, the experiment is referred to as a single factor experiment. In a factorial experiment, several factors may be varied simultaneously. In a factorial design the investigator selects a fixed number of levels (i.e. high and low) for each factor (i.e. shear, temperature, etc.) and then runs experiments with all possible combinations. For example, assume we are interested in the effect of shear, temperature, and humidity on a sensor's output. If we select a high and low level for each factor, then there are 2-levels$^{3\text{-factors}}$ = 8 combinations. In general, it can be shown that factorial experiments are the most efficient designs when the experimenter is interested in the effect of more than one factor. The following example illustrates some of the advantages of factorial design. Note that the example experiment is very simple and it is not very realistic. However, it is useful for illustration purposes.

Suppose we are interested in investigating the effect of temperature and shear on a pressure sensor. Using a single factor design a minimum of three experiments would be necessary. To obtain some degree of replication at least six experiments are necessary as shown in figure 11.8(*a*). Using a factorial design only four experiments are necessary to obtain the same information (see figure 11.7 and figure 11.8(*b*)). In fact the factorial design provides additional information. The factorial design also provides information on the interaction between two factors. Interaction is present when a change in one factor produces a different change in the response variable at one level of another factor than at other levels of this factor. In order to represent the interaction graphically we plot response vs temperature at both high (50) and low (10) levels of shear. In figure 11.9(*a*) we see that the shear lines are parallel, so there is no interaction. Figure 11.9(*b*) shows what the plot would look like if interaction were present. Note that in the factorial design no direct replications have been made. Because of this one might think that obtaining an estimate of error is not possible. Box *et al* (1978) discuss how it is usually possible to obtain estimates of error by making certain assumptions about interactions.

Temperature	Shear	Response (pressure)
–	–	71
+	–	67
–	+	85
+	+	82

Figure 11.7 The design matrix for a simple 2-level, 2-factor experiment. "–" indicates low level, and "+" indicates high level.

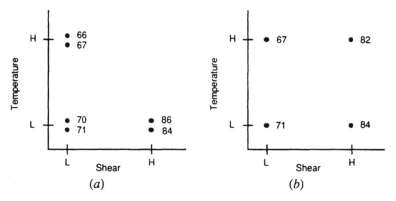

Figure 11.8 (*a*) Six trials are required to examine effects of shear and temperature using a single factor approach. (*b*) Using a factorial approach only four trials are needed. The factorial design provides additional information at high temperature and high shear.

The advantages of factorial designs are summarized in the following list:

(1) Factorial designs require relatively fewer runs per factor studied.
(2) Factorial designs provide information on possible interactions between factors.
(3) Factorial designs allow effects of a factor to be estimated at several levels of the other factors, providing conclusions that are valid over a range of experimental conditions.

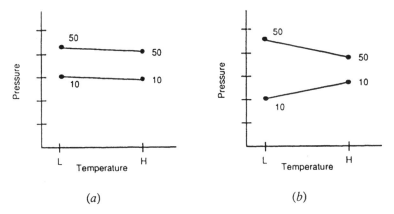

Figure 11.9 In both (a) and (b), the upper points labeled 50 represent the pressure output at the high shear level and the lower points labeled 10 represent the pressure output at the low shear level. (a) The shear response is similar (i.e. lines are parallel) at both levels of shear. So we can say that there is no interaction between temperature and shear. (b) Results differ at different shear levels so interaction between temperature and shear is present.

These advantages become more pronounced as the number of levels and/or factors is increased. This treatment of factorial designs is by no means complete. The reader is encouraged to consult the referenced material for a more rigorous treatment.

The goal of experimental design is twofold. First, we want to increase the amount and quality of the knowledge gained from the experiment. Second, we want to increase the efficiency of the experiment. In other words, through proper experimental design it is often possible to decrease the number of runs while at the same time increase the amount and quality of the information gained. Examining experimental design from a different perspective, we can state the goals in another way. In most experiments there are two important elements, a critical event and a perceptive observer. In the case of pressure sensors, the critical event might be the point at which the sensor saturates (i.e. the limits on the useful range of the sensor). The perceptive observer might be the experimenter or some instrument controled by the experimenter. In these terms, the goal of the experimental design is to bring these two elements together such that the perceptive observer is observing when the critical event occurs.

The steps to be followed when designing and analyzing an experiment are listed in the following procedure:

(1) Problem definition
   This step should not be overlooked. Careful, deliberate definition of the problem is essential.
(2) Data collection
   In this step, we decide on a method of data collection (static vs dynamic), the number of data points to be collected, the range over which the data is to be collected.

(3) Order and type of experiment

The order in which data will be collected and the method of randomization are determined. A choice is made between a variety of experiment types (single factor, factorial, etc.).

(4) Hypothesis

The experimenter defines the expected results in this step.

(5) Analysis

In this step, statistical methods are used to analyze the experimental data.

(6) Interpretation and presentation of results.

11.3 SHEAR MEASUREMENT

11.3.1 Importance of shear forces

The importance of shear in the formation of pressure sores is a widely discussed topic. First, what is shear? Fernandez (1987) uses the following definition: shear is when a force is exerted on the body, the skin remains in a fixed position, and the underlying tissues and skeletal system shift forward. The definition of shear is illustrated in figure 11.10. Section 2.1 discusses the difference between pinch shear and horizontal shear and the affects of both on blood flow. In this section we will concentrate on horizontal shear. The units of horizontal shear are pascals (N/m^2), where the force is applied horizontally, and the area is the area over which the shear is applied to the sensor. Bennett *et al* (1984) measured shear forces under seated patients that ranged up to 12 kPa.

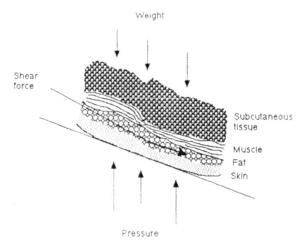

Figure 11.10 Gravitational force acting on an inclined support surface results in shear force. The skin remains stationary while the other tissue moves (Fernandez 1987).

Movement occurs when shear forces exceed frictional forces. Often an abrasion or blister will form as a result of rubbing due to movement.

Importance of shear in the formation of pressure sores
The role that shear plays in the formation of pressure sores is a controversial topic. A wide variety of theories exist. Guttman (1973) attributes a larger role to shear than to pressure in reducing the vascular supply. Reichel (1958) pointed to shear effects as a prime factor in sacral pressure sores developed in patients placed in reclining positions. Rudd (1962) believes that the kinking of vessels, when under the influence of shearing force, is a factor in deep necrosis. Roaf (1976) concludes that the avoidance of shear force is as important as the avoidance of direct pressure. Fernandez (1987) states that shear constricts and distorts the microcirculation already compressed by pressure. Unfortunately, largely due to the difficulty of accurately measuring shear forces, very little data exist to validity these theories. No commercially available sensors are adequate for measuring interface shear forces. The factors discussed in Section 11.2.1 apply to shear sensors as well as pressure sensors. An added difficulty when attempting to measure shear is the interaction of shear and pressure (Bennett *et al* 1979). Only the existence of large pressure values permits the stable development of large shear values. Most shear sensors are based on the deflection of a deformable element. Because of this it is often hard to eliminate the effect of pressure on the deformable element. This problem is addressed in more detail in Section 11.3.2.

Two studies have produced experimental evidence which supports the theory that shear forces do play an important role in the formation of pressure sores. When applying pressure and shear to swine, Dinsdale (1974) found that the combination of pressure and shear caused ulceration with pressure as low as 45 mm Hg. When applying pressure alone, 290 mm Hg was required to produce ulcers. Bennett *et al* (1979) examined the effect of shear in reducing the pulsatile arteriolar blood flow. They showed that the magnitude of pressure necessary to produce occlusion can be nearly halved when accompanied by sufficient shear.

11.3.2 Interaction of pressure and shear

To illustrate the interaction of pressure and shear on a shear sensor, let us examine a shear sensor built and tested by Bennett *et al* (1979). The sensor consists of a dual cantilever beam surrounded by an essentially immovable block of metal (figure 11.11). To sense shear the cantilever beam must deflect in the direction of the applied shear force. This deflection is then sensed by the strain gages mounted at the base of each cantilever beam. With some applied pressure but no applied shear the shear sensor should provide a zero output. As shear is applied the cantilever beams begin to deflect in the direction of the shear and provide an increasing output. The applied pressure will tend to increase this deflection giving rise to a pressure-upon-shear error. In general the smaller the deflection, the smaller the error.

Motion error
Note that the immovable metal block shares the same soft tissue loads as the shear sensor. If we assume that the skin does not slip with respect to either the

block or the shear sensor, then any motion of the shear gage gives rise to some relative compression in that section of tissue over the gap between the block and the advancing sensor. In a similar manner, additional tension is created in the section of tissue over the gap between the block and the retreating shear sensor. Both the tension and the compression created will act to resist the motion of the shear sensor. As a result the shear sensor will produce shear values less than the true value of the local shear. This error is often referred to as motion error. Any shear sensor which is based on a deformable element is subject to some degree of motion error. Usually if the amount of deflection or deformation can be made very small the motion error can be minimized. Unfortunately, reducing the deflection usually reduces the sensitivity of the sensor as well. For example in the cantilever beam sensor, reduced deflection implies a stiffer cantilever beam (i.e. reduced sensitivity).

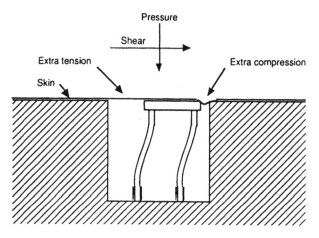

Figure 11.11 A cantilever beam shear sensor consisting of two attached beams instrumented with strain gages at the base. Since the skin is essentially fixed to the sensor and the surrounding block, tension and compression of the skin occur in the gaps (adapted from Bennett *et al* 1979).

In order to test the motion error, Bennett *et al* (1979) used the setup shown in figure 11.3. To simulate tissue a piece of reinforced silicone gel was placed on the sensor head. A 500-g brass weight was used to develop sufficient normal pressure to eliminate slip under a simulated shear load. The shear load was delivered through a light thread. By comparing the area of the brass weight base with that of the shear sensor and assuming that the sensor absorbs its proportional share of the total shear load, it was possible to test the accuracy of the shear gage under low tissue stress conditions.

Other sources of motion error

In addition the basic stress level existent in the tissue over the sensor affects the amount of motion error. If the stress level is high, the introduction of a further increment of shear, even if small, will lead to a large increment of motion error. In the same manner, if the stress level is low, even large shear increments will produce small motion errors. In other words, if the tissue over the sensor were perfectly elastic no motion error would be present. So the higher the stress level the less elastic the tissue will be. Also, different skin types may lead to varying amounts of motion error. The skin of an elderly person may tend to be less elastic that that of a young person. Prolonged exposure to sunlight gives the skin a leathery, less elastic characteristic. However since we are interested in measuring shear in areas prone to the development of pressure sores and these areas are generally protected from exposure, the skin will remain elastic just like a baby's. Finally, the location of the sensor can produce different levels of stress. The skin over a bony prominence will tend to have a higher stress level which may lead to a significant motion error whereas testing over a muscle will lead to smaller errors.

11.3.3 Methods of measuring shear forces

In this section we will present a number of sensors for measuring shear forces. Some of the methods have been built and evaluated, some exist on paper only. All of the sensors are based on some type of deformable element. The sensor is designed so that the applied shear force will cause the deformation. The amount of deformation or deflection of the element is then measured using a variety of sensors. Finally, the output from the sensor can then be related to a shear force.

Bennett et al (1979) cantilever
Bennett *et al* (1979) developed an instrument capable of measuring horizontal shear, pressure, and pulsatile arteriolar blood flow. Figure 11.11 shows the shear part of the sensor, which employs a combination of cantilever beams and strain gages. The total sensor consists of four sensors (2 pressure, 1 shear, 1 blood flow). The sensors were placed within a 25-mm circle and embedded in a larger (77-mm diameter) aluminum block in order to present a flush surface. Figure 11.12 shows this configuration. Thus the individual sensors monitor slightly different, albeit adjacent, portions of tissue. Section 11.3.2 discusses in detail the effects of the interaction of pressure and shear on this sensor. The application of the maximal pressure encountered to the shear sensor at maximal displacement produced no measurable change in shear output, thus the pressure-

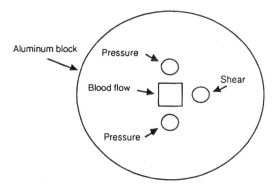

Figure 11.12 Top view of pressure/shear/blood flow sensor. Diameter of aluminum block was 77 mm. The entire sensor was flush-mounted within a wheelchair seat. To the subject the individual sensors appeared to be flush-mounted buttons (Bennett *et al* 1979).

upon-shear error was negligible. This is probably due to the small (25-μm) deflection of the shear sensor at maximal shear. The motion error was shown to be less than 5% at low shear values. Although the sensor worked well as used by Bennett *et al* (1979), it is not suited to measuring interface shear forces between compliant surfaces. The sensor is rigid and hard; the hospital bed is soft and compliant. To introduce a hard gage into a soft environment is to generate false readings (Bennett 1990).

Silicone elastomer-encased cantilever

We have built a prototype silicone elastomer encased cantilever beam as shown in figure 11.13. The cantilever beam is made of beryllium-copper. Initially we attempted to make the cantilever of spring steel, but the spring steel was difficult to shape. First the steel was heated until red hot and then allowed to cool. This removed the temper and enabled the steel to be shaped without breakage. After the steel was shaped it was again heated red hot and then quickly quenched in water. After several attempts, we discovered that the steel became too brittle after quenching. The solution was to draw back the temper by annealing the steel at a lower temperature. Because the beryllium-copper is much easier to work with we chose to pursue it instead of the spring steel. After forming the cantilever beam out of beryllium-copper, it was riveted to an aluminum base. A rigid base is required to yield deformation due to pressure. Thus the cantilever beam itself is the only deformable part of the sensor. Two metal strain gages were then mounted near the base of the cantilever beam.

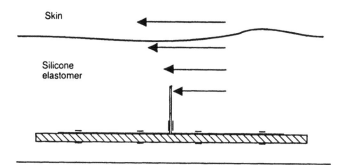

Figure 11.13 A beryllium-copper cantilever beam is mounted on an aluminum base and encased in silicone elastomer. Shear forces are transmitted from the skin through the silicone elastomer to the cantilever beam.

The unpackaged sensor was calibrated by hanging weights on the end of the cantilever. The sensor output was essentially linear, with some hysteresis. When the cantilever was under full load, we quickly removed the entire load. The output immediately dropped to within 5% of the original unloaded output. Then over a period of about 5 min the output gradually returned to the original unloaded output. We believe this effect is due to frictional forces between the two cantilever beams. Once the load is greater than about 5% (20 g), the effect is negligible because the applied force is large enough to overcome the frictional force.

Next, we encased the entire sensor in 10-mm thick silicone elastomer. The cantilever was covered with silicone elastomer and cured at 70°C for 90 min. The encased sensor was tested with a force of 58 kg (80 kPa) applied to the top of the sensor. Shear was applied using a spring scale up to a maximum of 40 kPa. We expected the output to be linear, especially at low levels of shear. At higher levels of shear, we expected some nonlinear behavior due to the pressure-on-shear interaction. The experimental results were quite linear. Some evidence of the pressure-on-shear effect was present at higher levels of shear. However, more testing is necessary to verify this effect.

Thus we demonstrated the feasibility of using a silicone-encased cantilever to measure shear forces under the sacrum in elevated hospital beds. By minimizing the size of the cantilever and encasing an array of sensors in a single rubber pad shear distributions could be measured.

Box design

A combination pressure/shear sensor could be constructed as shown in figure 11.14. The sensor is based on an open-ended box made of a suitable metal. The top and bottom of the box should be much stiffer than the sides in order to minimize bending in the top and bottom from applied pressure. The sides of the box would act as cantilever beams. Strain gages attached to the sides would measure the horizontal deflection due to shear. A simple conductive polymer pressure sensor could be mounted on the top surface to monitor pressure. The height of the box should be as small as possible, while the length and width should be chosen to provide the desired resolution and sensor performance. One advantage of this design would be its ability to measure pressure and shear at one location simultaneously.

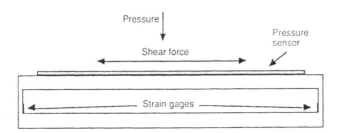

Figure 11.14 Strain gages on the ends of an open-ended box could measure shear, while a pressure sensor on the top could measure pressure.

Pneumatic design

Lewis and Nourse (1978) presented one of the most novel shear sensor designs, which measures deformation using pneumatics. The basic operating principle of the device is the measurement of the relative motion between a small area of skin surface and the larger surrounding area of skin. Figure 11.15 shows that a small disk, in firm contact with the skin surface, mounted on a steel wire can move relative to the surrounding portion of the device which is in firm contact with the larger surrounding area of skin as well as with the bedsheet or other support material. The steel wire also acts as a spring to return the disk to a central position. The disk is also supported on Teflon riders to reduce the friction of the system. Measurement of the deflection of the disk is accomplished by means of four pneumatic flapper valves set 90° apart from one another. A flapper valve is really a pair of valves in which the motion of the disk under the influence of shear forces is toward one nozzle while simultaneously being away from a second nozzle. Opposing valves are coupled to provide a pressure differential output. By using two pairs of flapper valves arranged in a mutually perpendicular manner, the direction of the applied shear could be detected. Overall dimensions of the device are approximately $23 \times 23 \times 3$ mm. The central detecting disk is 6 mm in diameter. A can of Freon gas is used to supply air to the sensor. As with Bennett's cantilever beam sensor, motion error as described in Section 11.3.2 is a potential problem with this sensor.

Fiber optic design
Section 9.6 describes a fiber optic pressure sensor. Figure 11.16 shows modifications made to the pressure sensor design to convert it to a shear sensor. As horizontal shear is applied to the top rubber surface, the reflecting surface moves in the direction of the applied shear. The moveable section is mounted on rollers or Teflon blocks in order to minimize the effect of pressure (i.e. the resistance to horizontal movement should be independent of normal pressure). The horizontal movement is resisted by a spring at each end. The reflecting surface is graduated (see figure 11.16(b)). As a result, as the amount of shear changes the amount of reflected light will change. If the characteristics of the springs are known, then the change in the light output intensity can be related to the applied shear force.

Other designs
A capacitive shear sensor is also feasible. A design could utilize a simple parallel plate capacitor where the top plate could slide horizontally under shear. The sliding top plate would then cover a smaller area of the bottom plate and thus change the area of the capacitor. The resulting change is capacitance could then be related to the applied shear force. Fan *et al* (1984) describe a capacitive shear-sensitive tactile sensor.

Cooper *et al* (1986) used simple Hall-effect sensors to monitor shear forces. However, the sensors are quite large and not suitable to measuring interface shear forces.

11.4 STUDY QUESTIONS

11.1   Sketch the structure of a manually actuated pressure calibrator and its data output channels.

11.2   Sketch the structure of a pneumatically actuated pressure calibrator and its data output channels.

11.3   Describe the steps carried out when performing regression analysis on calibration data.

11.4   With the aid of a pressure distribution diagram, explain how sensor size can cause error in pressure measurement.

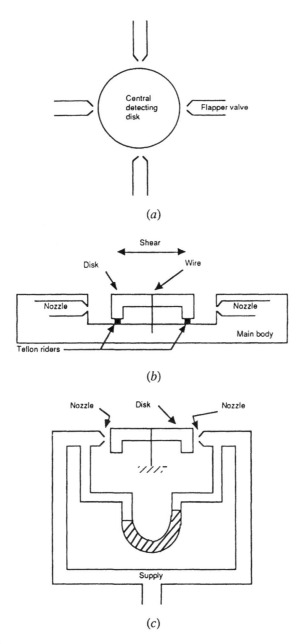

Figure 11.15 (a) Top view of central detecting disk with four flapper valves (nozzle). (b) A central detecting disk is supported by a single wire that acts as a spring to return the disk to a central position. (c) As shear is applied to the central disk it moves toward one nozzle and away from the opposite nozzle. This deflection causes the pressure in the chambers behind the nozzles to change. This change can then be related to the applied shear force (Lewis and Nourse 1978).

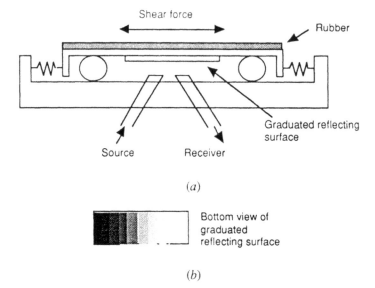

(a)

Bottom view of
graduated
reflecting surface

(b)

Figure 11.16 (a) A rubber surface provides high friction so shear force moves the moveable plate over low-friction ball bearings to compress the spring. (b) The graduated reflective surface produces a light intensity output that varies with the position of the moveable plate.

11.5   With the aid of a pressure distribution diagram, explain how load shape can cause error in pressure measurement.

11.6   Describe how poor experimental design can produce experimental errors.

11.7   Explain why factorial experiments usually preferred over single factor experiments.

11.8   Explain the evidence that shear is an important factor in the formation of pressure sores.

11.9   What is motion error and how does it affect shear measurements?

11.10   Sketch and explain the principle of operation of a cantilever beam shear sensor.

11.11   Sketch and explain the principle of operation of a box design shear sensor.

11.12   Sketch and explain the principle of operation of a rubber encased shear sensor.

11.13   Sketch and explain the principle of operation of a capacitive shear sensor.

11.14   Sketch and explain the principle of operation of a fiber-optic shear sensor.

12

Behavior for Relieving Pressure

Mohammad R Akbarzadeh

This chapter first describes different pressure–duration curves to show that relieving pressure is necessary to prevent pressure sores. Individuals with normal skin sensation and enough strength to shift their body weights avoid pressure sores by frequent body movements. For those without normal sensation, it is desirable to develop pressure relief behaviors as a permanent habit. We describe different methods for monitoring and training behaviors for pressure relief. The last section describes different possibilities for socially acceptable alarms to be used in training devices.

12.1 RELATION BETWEEN PRESSURE AND DURATION

It is important to know the relation between the amplitude and duration of pressures applied to the skin that cause pressure sores. This information can help us investigate whether the pressure exceeding mean capillary pressure (32 mm Hg) is the main cause of pressure sores or whether other factors are more important. It can also be a good guideline to determine the minimal frequency for pressure reliefs where the pressure is the dominating factor for producing pressure sores.

12.1.1 Review of related works

Groth (1942) applied different pressures to the posterior ischial tuberosities of rabbits via two 15 mm circular disks. He observed that most severe pathologic changes occurred in deep muscle while there were small changes in the skin. He reported an inverse relation between the pressure and duration to produce pressure sores. In his experiments, paralyzed and control animals behaved the same in producing pressure necrosis.

Kosiak (1959) applied pressures ranging from 100 to 550 mm Hg to 16 male and female mongrel dogs. The apparatus to apply the pressure consisted of the plunger of a 20-ml hypodermic syringe inverted and stabilized over the test area, a compressed air tank with a pressure reduction valve, and a mercury manometer. To measure the effective pressure applied to the tissue, he also used a 20-gage needle inserted into the tissue. The needle was connected to a mercury manometer for monitoring and a strain gage for recording the pressure. A constant injection pump infused the saline into the tissue at the rate of 4.6 ml per hour to assure the contiguity between fluids. In 62 separate experiments, he observed and plotted the inverse relation between minimal pressure and duration parameters for the production of pressure sores (figure 12.1). He reported that tissues subjected to pressures as little as 60 mm Hg for only 1 h showed microscopic pathologic changes, however his pressure–duration curve showed that minimal pressure for producing the ischemic ulceration was 180 mm Hg applied for 6 h. His pressure–duration curve shows that applying a pressure of 100 mm Hg for 12 h did not cause any ulceration. He observed that all the tissue from the skin to the bone was under sufficient pressure for simultaneous degeneration. He also reported that complete pressure relief might often prevent ulceration even though excessive pressures were applied for a sufficient duration.

Kosiak (1961) performed 80 separate experiments on rats to determine the effect of both constant and alternating pressure applied to the muscle of normal and denervated rats. The apparatus that applied and measured the pressure was the same as in his previous report on dogs. He observed the inverse relation between intensities and durations of pressure necessary to produce pressure sores. There were not any detectable differences between normal and denervated muscles in developing microscopic changes. He reported that constant pressure as low as 70 mm Hg applied for a relatively short duration (2 h) could produce microscopic changes in the muscle. Microscopic changes in the muscle were absent or less prominent for alternating pressure application.

Reswick and Rogers (1976) plotted allowable pressure vs duration based on over 980 clinical observations (figure 12.1). They obtained their data from actual patient experience in the following ways: subjective comments by physicians, nurses, and therapists on pressure sore development, actual pressure and time measurements when the patient's skin showed signs of potential or actual development of pressure sores, and a few controled tests on volunteers where they applied pressures for times sufficient to produce clinical signs of potential breakdown. Considering the differences in patient's skin and general health and the relatively a few controled measurements, they stated that their pressure–duration curve could give general guidelines and should not be taken as absolute. This curve presents average data from a select group of patients, therefore it is of little use in predicting the response of a given individual to pressure.

Daniel et al (1981) used a continuously monitored computer-controlled electromechanical pressure applicator to ensure that the amount of pressure applied to the skin remained constant during the experiment. They applied pressures ranging from 30 to 1000 mm Hg for different durations on 30 pigs. They found that initial pathologic changes were in the muscle and could progress toward the skin with increasing pressure and/or duration. They obtained the pressure–duration curve shown in figure 12.1. Squares are for occurrence of muscle damage and triangles are for muscle and skin damage.

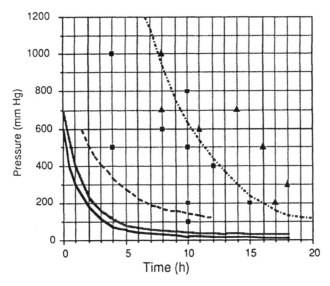

Figure 12.1 Inverse relation between pressure and duration for producing pressure sores reported by different investigators (Adapted from Reswick and Rogers 1976, Kosiak 1961, and Daniel *et al* 1981).

════════	Range of maximal acceptable pressure/time application over bony prominences suggested by Reswick and Rogers (1976)
– – – – – –	Minimal pressure/time application to produce pressure sores in dogs experienced by Kosiak (1961)
∙∙∙∙∙∙∙∙∙∙∙∙	A critical pressure–duration curve for producing pressure sores in normal pigs experienced by Daniel *et al* (1981)
■	Muscle damage reported by Daniel *et al* (1981)
▲	Muscle and skin damage reported by Daniel *et al* (1981)

12.1.2 Analyzing pressure–duration curves

Daniel *et al* (1981) pointed out some of the factors which must be considered if we want to use the results of pressure–duration curves. To extend the results to the human, the experimental animal should have the same skin characteristics as the human. Animals may be classified based on skin characteristics into two groups: loose-skin and fixed-skin. For example, the dog is a loose-skin animal and does not have comparable subcutaneous tissue to cushion the pressure load. Like humans, the pig is a fixed-skin animal and has thick, well vascularized, and rigidly adherent skin. Therefore, the pig is the preferred experimental animal (Daniel *et al* 1981). For pressure–duration studies in paraplegic animals, sufficient time (6 weeks) should pass the denervation to permit weight gain and extensive atrophy of muscle and subcutaneous tissue. Another important factor is the accuracy and reliability of pressure measurements. Moreover, soft tissues compress with time causing variations in the applied pressure. Considering these factors, Daniel *et al* (1981) chose the pig as the experimental animal and used a computer to continuously adjust a stepping motor to apply constant pressure on the skin. The large difference between pressure–duration curves for experimental animals in figure 12.1 might be due to the different skin characteristics of the dog and pig and also different methods used for pressure measurements. Considering muscle damage presented by the squares in figure 12.1, the safe level for pressure–duration profiles of two separate experiments on animals would be closer to each other. Section 12.2.2 shows which curve might be more applicable to the human to determine minimal push-up frequency.

Pressure–duration profiles from experimental animals reported by both Kosiak (1959) and Daniel *et al* (1981) are much greater than the curve from clinical observations for the human obtained by Reswick and Rogers (1976). This difference might be due to other pressure sore producing factors in clinical situations such as shear forces and dynamic loading during daily activities. To compensate for these factors, the safe pressure–duration profile must be much lower (Patterson 1984). Another factor is the role of bacteria in pressure sore production. Presence of bacteria at the site of pressure application results in breakdown of tissue with lower pressure (Groth 1942). Bacteria may play a significant role via secondary infection from endogenous sources, mainly the urinary tract (Daniel *et al* 1981). This might explain different answers to the question: "Is the origin of pressure sore development in the skin or muscle?" Models for soft tissues in Chapter 2 showed that pressure should be higher near the bony prominences than at the surface which explains why in animal experiments ulceration starts from the muscles and then spreads to the skin. However, in real situations other factors such as shear forces, dynamic loading, moisture, and bacteria make the skin more vulnerable for ulceration. Since these factors vary from one person to another, acceptable limits for pressure–duration are different.

12.2 WHEELCHAIR PRESSURE RELIEFS

Mooney *et al* (1971) measured the pressures under the ischial tuberosities of individuals sitting on different seat cushions. Pressures ranged between 51 to 86 mm Hg for normals and 83 to 152 mm Hg for SCI persons. These pressures are much higher than the pressure in capillaries and if sustained on the skin for an extended duration they can cause pressure sores. Comparing these results with pressure–duration relations discussed in the previous section shows that persons using wheelchairs need to develop the habit of relieving pressure under their ischial tuberosities frequently to prevent pressure sores. Paraplegic persons who have strong triceps are instructed to perform push-ups regularly while seated in their wheelchairs. A conventional push-up is to place each hand on the wheelchair armrests or wheels and extend both elbows to approximately 180° for 3 s or more which results in the buttocks lifting off the cushion (White *et al* 1989). Tetraplegic persons cannot perform push-ups but some of them can partially shift their weight to relieve the pressure under their ischial tuberosities by forward or side leans. Griffith (1963) suggested that paraplegic persons should perform push-ups every 30 min and lie down for 15 min every 2 h. There are no standard frequencies and durations for push-ups but it is clear that more frequent and longer push-ups are more effective in preventing pressure sores. The recommended intervals between push-ups range from 15 min to 2 h and the duration is at least 3 s up to 30 s. In current clinical practice 15–30 min interval and 3–5 s duration is prescribed (Grip and Merbitz 1986). Merbitz *et al* (1985b) instruct their subjects to perform push-ups every 20 min for at least 3 s.

12.2.1 Techniques used to monitor wheelchair pressure reliefs

Many investigators have developed different techniques to monitor pressure relief behavior of paraplegic and tetraplegic persons. These measurements can serve the following purposes: (1) examining the relation between pressure relief behavior and pressure sores, (2) training paraplegic and tetraplegic persons to develop a permanent habit of pressure relief, (3) evaluating the effectiveness of training programs, and (4) identifying SCI wheelchair users who are at risk of pressure sore development. The following paragraphs describe some of the techniques developed for monitoring pressure relief behavior.

Patterson and Fisher (1980) taped 1 mm thick and 5 mm in diameter Entran Model ESP–200 pressure sensors under each ischial tuberosity. They calibrated sensors in a small pressure chamber with a mercury manometer for reference. They determined the zero pressure reference by averaging the values observed at the start and at the end of the experiment. A 4-channel tape recorder was used to store data. They used this technique to record pressure and push-up patterns for paraplegic patients to investigate the relation between push-ups and pressure sores. The average recording time was 5.5 h. The advantage of this technique is the direct measurement of pressure under the tissue and the disadvantage is difficulties and inconvenience in placing sensors. The purpose of this study was to measure the variation of pressure under the ischial tuberosities of paraplegic patients. Fisher and Patterson (1983) used the same technique to monitor sitting behavior of tetraplegic patients.

Jaeger and Marlantis (1983) replaced the original wheelchair armrest with a modified one which had strain gage pressure sensors and all other circuitry. The device could monitor the frequency and duration of push-ups and warn the subject by turning on an alarm if no push-up with sufficient length was detected in a preset time interval. The advantages of this design over others are its small size and inconspicuous mounting as a part of normal wheelchair arms. It is very suitable for folding wheelchairs.

Merbitz *et al* (1985a, 1985b) used a large air-tight bladder placed under the subject's seat cushion. A tube connected the large bladder to a small vinyl bag which was located in a box and could determine the position of a switch based on the weight on the cushion. A battery-operated microprocessor in the box stored the position of the switch at fixed time intervals, therefore they could monitor daily sitting behavior of the subject by plotting the stored data. This technique is easy to use for monitoring push-ups but does not provide details about pressure variations. They used this device to train and evaluate pressure relief behavior.

Burn *et al* (1985) designed a microprocessor-based pressure relief monitoring device which can operate for 30 days without intervention. The device consisted of three components: force sensors, detecting circuitry, and monitoring computer. When the subject made a movement, if the magnitude of the movement exceeded the preset threshold, the detecting circuitry would turn the monitoring computer on to evaluate and record the movement. The evaluation of the movement was based on the calibration performed at the time of device installation. The force sensor had four load cells and each load cell had four semiconductor strain gages in a full bridge configuration.

12.2.2 Relation between pressure relief and pressure sore development

Patterson and Fisher (1980) investigated the pressure relief patterns under the ischial tuberosities of 12 paraplegic subjects during their daily activities using the technique described in the previous section. Subjects sat at pressures greater than 30, 90, and 150 mm Hg for 91.8 ± 19.0, 53.5 ± 34.8, and 17.6 ± 10.7% of the time respectively. The average sitting time without a push-up for durations greater than 1 and 5 s was 10.1 ± 6.4 and 29.6 ± 27.5 min respectively which was in acceptable limits for ulcer prevention (10 to 30 min). Two of the subjects experienced frequent uninterrupted pressures for a long period of time (greater than 60 min) without any push-ups but did not have pressure sores.

Fisher and Patterson (1983) recorded pressures under the ischial tuberosities of five tetraplegic subjects. The sitting behavior for each subject remained relatively unchanged. The average measured pressures were 71.5 and 105.4 mm Hg on the ROHO and foam cushions respectively. The average time between push-ups greater than 1 and 5 s was 72.1 and 96.2 min. These average values were much greater than those usually recommended for skin care but none of the subjects developed pressure sores. These results confirmed the study of Newson *et al* (1981) for applying pressures for long periods on healthy skin without causing pressure sores. Fisher and Patterson (1983) concluded that pressures at which the partial pressure of oxygen became zero had to be higher than the generally considered pressure limit for tissue viability. They also considered that frequent uninterrupted pressure could lead to progressive damage of muscle and later of skin. They compared the results from this study with the previous study on paraplegic patients and stated that tetraplegic patients sat longer between push-ups and had less pressure oscillations than paraplegic subjects did but they never developed pressure sores. They concluded that beside pressure, shear and friction forces in paraplegic patients are important factors in developing pressure sores.

Patterson (1984) showed that in many cases average pressure and time values under ischial tuberosities of paraplegic and tetraplegic subjects were inside the unsafe region of Kosiak's pressure–duration curve and those subjects did not have pressure sores within a six-month period of the experiment. Those pressure and duration values could be mapped inside the safe region of Daniel's pressure–duration curve. From these observations, we might conclude that Daniel's pressure–duration curve is more applicable to the human where pressure is the dominant factor for producing pressure sores.

Merbitz *et al* (1985b) found wide variability in the sitting behavior between patients and within patients over time. They concluded that there was no simple relation between push-up intervals and development of pressure sores and prolonged sitting without push-ups was not a sufficient cause of pressure sore development. Merbitz *et al* (1985a) observed pressure sore formation in two individuals who did not relieve pressure for extended periods. Considering all these results, Grip and Merbitz (1986) hypothesized that pressure relief requirements may be different for each individual and may change as other related factors vary.

12.2.3 Training pressure relief behavior on the wheelchair

It is a challenge to develop pressure relief behavior as a permanent habit in paraplegic and tetraplegic patients. The first step is educating patients about pressure sores, factors governing them, and requirements for pressure relief. The second step is to train the individual to practice good hygiene, take precautions to reduce friction and shear, and perform acceptable pressure reliefs at the correct intervals. The nursing staff should monitor the effectiveness of the training program by regular inspection of the skin. Patients should maintain this self-care practice by frequent pressure reliefs and daily skin inspection (Grip and Merbitz 1986). The role of technological developments is to help the patient develop a permanent habit to relieve pressure and also to assist the patient and the nursing staff in evaluating the effectiveness of the training program. Technological developments can also be used for identifying patients at high risk of pressure sore development.

An old and simple device used for pressure relief training is a mechanical timer which cues patients to relieve pressure periodically. It needs to be rewound each time which might be impossible for many quadriplegic persons. This type of timer is an obvious symbol of disability. Klein and Fowler (1981) used a commercially available timer/microcalculator which emitted high-frequency tones at the end of a specified time interval. They believed that it was an effective and economical approach for training SCI patients in and out of the hospital.

Monitoring devices can help rehabilitation professionals train and monitor their patients' performance. These devices are for biofeedback applications which are based on operant or instrumental learning or conditioning. In a biofeedback system, a controller monitors the physiological or behavioral response of a subject and produces proper stimuli to condition the response. A typical example of this behavioral conditioning is a food- or water-starved animal which receives a small amount of the withheld substance as reward for correct responses. Figure 12.2 shows a typical block diagram of a microprocessor-based training device. The device monitors sitting behavior of the subject which can be the frequency and duration of push-ups or pressure distributions under ischial tuberosities. Based on a preprogrammed algorithm, the controller produces appropriate feedbacks to condition the behavior and develop a push-up habit. Different types of alarms and feedbacks are discussed later in Section 12.4. The following paragraphs describe different examples for each block shown in figure 12.2.

Patterson and Stradal (1973) developed a device that emitted a warning when the patient remained seated for 10 min. They placed a pressure-sensitive switch under the seat cushion. The device was automatically reset when the seat pressure was removed for 4 s. However, they found that patients would not accept any warning device.

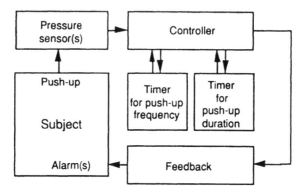

Figure 12.2 Block diagram for a pressure relief training device based on the concept of biofeedback.

Malament *et al* (1975) developed a training and monitoring device to improve pressure relief training. They used the principle of negative reinforcement in which the number of responses that stops a negative stimulus (in this case a 30-s buzzer) will increase. Their system had a pressure mat placed on the seat of the wheelchair and could only transfer relief events to the monitor. The patient could avoid the alarm by performing push-ups every 10 min which resulted in resetting the 10-min timer. The device was capable of recording the number of push-ups and alarms (10-min intervals without a push-up). They monitored sitting behavior of five SCI patients in three phases. In phase 1, there was no feedback from device so they could compare the improvement in the final phase. In phase 2, the alarm was activated and patients were instructed to try to prevent the alarm from going off by performing push-ups every 10 min. For phase 3, the alarm was deactivated without the patient's knowledge. Two weeks after phase 3, in follow-up evaluation, they monitored patients as in phase 3 to see whether the push-up behavior persisted for a longer period of time. Sitting behavior of patients was evaluated by comparing the number of push-ups and alarms in each phase. The results from this study showed the effectiveness of this system to train and evaluate the sitting behavior of paraplegic patients. They pointed out that instructions alone were not enough to develop push-up behavior in phase 1 and using the alarm in phase 2 gradually increased the number of push-ups. It took several days to develop the habit but it persisted after the alarm was disconnected in phase 3 and in the follow-up evaluation.

One important factor in evaluating training devices is the Hawthorne effect. People may behave differently when they know they are experimental subjects and being monitored. To reduce this effect, it is advantageous to have a device which can operate for a long period of time without intervention. It is also desirable that the training device can be used as an evaluating system with minimal intervention that attracts the subject's attention.

Chawla *et al* (1978–79) developed a warning device similar to the previously described system for push-up training. In phase 2, there were two types of feedbacks for warning the patient. A light turned on as a reminder for pressure relief when the patient sat for 10 min without performing a valid push-up with a duration of 20 s. An audible alarm was used if the patient failed to perform a valid push-up following the light alarm. The device recorded the number of times each alarm was activated and also the time of push-ups. From the tabulated results, the authors came to the following conclusions. In phase 1, in spite of verbal instructions, patients had a tendency not to relieve pressure regularly. Using the training system and warning alarms, in phase 2, had an immediate effect in forming a regular sitting behavior. Monitoring the patients in phase 3 showed that the intervals between push-ups were irregular but the number and duration of push-ups remained steady. The authors felt that audiovisual reminders might help to improve the pressure awareness of the patient and they were in the process of developing a device which would remind the patient to relieve pressure and could be given to him when discharged from hospital.

Cumming et al (1986) used a microprocessor–based weight shift monitor for training paraplegic patients to develop the push-up habit. An audible alarm turned on whenever patients failed to perform a valid push-up in a preset time interval. The alarm could be disabled to evaluate the patient's self-timing or for certain social situations. The device stored sitting behavior of the patients (frequency and duration of push-ups) in its memory. They could transfer data to a microcomputer to evaluate the patient's performance.

White et al (1989) used the same monitoring technique developed by Merbitz et al (1985b) to evaluate the push-up training of two 11-year-old subjects with spina bifida which results from birth with an open spine. They used a watch programmed to beep every 30 min and stimulate a push-up. If the subject failed to perform a push-up of at least 3 s, the microprocessor-based device located under the participant's wheelchair seat would emit an unavoidable pulsating 86-dB alarm for 6 s (alarm avoidance). The device recorded the number of appropriate 3-s push-ups for varying conditions for more than 2 months:

(1) Subjects were instructed to perform push-ups while watch beeper and alarm were disabled.
(2) Both watch beeper and alarm avoidance were enabled.
(3) Only alarm avoidance could sound if an appropriate push-up did not occur within 30 min after the last push-up.
(4) Only watch beeper prompted push-ups.

For 1 more month, procedures for condition 2 were continued without data collection and follow-up data were collected for the next 3 days. The results from this experiment suggested that the combination of instructions, prompts, and avoidance procedure was effective in developing and maintaining pressure relief behavior. The number of appropriate intervals (10-min interval with a valid push-up) doubled by using both feedbacks. Examining the relative effectiveness of various intervention components (beeper and alarm avoidance) showed that eliminating one of them (beeper or alarm) decreased the number of appropriate intervals (by 50% or 66% respectively).

Webster (1989) proposed a monitoring/training device in which a pressure pad containing an array of 64 Interlink conductive polymer pressure sensors would be placed between the wheelchair cushion and patient to obtain a close measurement of the pressure distribution on the skin. The device would alarm whenever the pressure–duration distribution on any sensor exceeded a preset acceptable level. The safe level would be based on the pressure–duration curve developed by Reswick and Rogers (1976). This device would provide more information than previously described devices by measuring the pressure–duration distribution over a large area and would only warn the patient whenever a pressure relief was really necessary and not at fixed intervals as other simple devices did. Potential difficulties in building such a device would be choosing thin elastic compliant-enough materials for the pad, placing and maintaining calibration for pressure sensors, arranging wire connections for the compliant pad, high cost, and obtaining user compliance. Another problem would be to determine the pressure–duration curve for each individual, which presumably would vary from that of Reswick and Rogers.

12.3 PRESSURE RELIEFS DURING SLEEP

There are different purposes for studying body movements, especially spontaneous movements during sleep. For sleep studies, body movements during sleep might be related to quality of sleep and other factors affecting sleep. Body movements of subjects on different mattresses might help us evaluate their effectiveness in relieving pressures. Measuring body movements of elderly patients and scoring their mobility can also help us identify patients at high risk for pressure sore development. Measuring average time between postural changes in normal subjects might be useful in deciding about the minimal physiological mobility requirement in nursing disabled patients.

12.3.1 Factors affecting body movements during sleep

Kleitman (1939) reviewed many studies on body movements during sleep and different factors governing body movements. The frequency and distribution of movements are discussed later in this section. Figure 12.3 summarizes the effects of different factors considered by those studies. Body movements during sleep for each individual depend on personal characteristics and may change from time to time. In studying sleeping behavior for pressure sore scoring or testing different mattresses, factors listed in the table and other possible factors should be controled very carefully.

12.3.2 Monitoring body movements during sleep

Direct observation of patients' behavior during sleep is not practical especially for studies of many subjects. Investigators have developed different types of devices which can detect and count different movements. Kleitman (1939) described some monitoring techniques which were combinations of mechanical devices for detecting and counting movements. Advances in electronics have yielded less costly and more efficient devices for recording body movements. Exton-Smith and Sherwin (1961) developed a device to record movement of patients in bed. The device consisted of an actuating assembly, an inertial switch, and an electronic counter. Muzet *et al* (1972) used three bedsteads connected to displacement transducers to detect subjects' movements in any three possible directions. They used logic circuits to trigger the camera shutter to create a pictorial record of the subject's postures and positions during sleep. By using a pair of load cells placed under the bed legs, Bardsley *et al* (1976) could identify types of movements. For each movement, the output signal from a load cell had two components: a transient due to dynamic forces generated by the body and a steady state component representing the position of the center of gravity of the body. Load cells were calibrated for each subject performing different position changes. Output signals from two load cells could be used to identify the nature of movements. For a pictorial record, major movements could trigger the camera shutter and infrared flash. Wheatley *et al* (1980) designed low-profile load cells to be placed under bed legs for monitoring movement during sleep. These cells were designed to be responsive only to uniaxial forces. Since hospital beds usually have wheels, load cells had a large surface area for different wheel sizes and relatively independent response for different wheel positions. Webster (1989) suggested the use of an array of Interlink pressure sensors mounted on a thin compliant pad to measure the pressure–time distribution on the skin and warn the nursing staff to turn the patient whenever it is necessary. Advantages and disadvantages of this proposal were discussed in the previous section.

Factor	Relation to body movements
Percentage of CO_2 in alveolar air	Inversely affected the frequency
Body temperature (also fever)	Directly affected the frequency
Temperature difference between skin and bed (normally 4 to 5°C)	Increased when the difference was reduced to 1°C
Emotional states	Stress or shock could cause more body movements
Large meal before sleep	Increased body movements
Alcohol	No effects on average frequency but reduced body movements during the first half and increased restless during the second
Pressure of urinary bladder	Directly affected the frequency
Going to bed early	Reduced body movements
Taking hot or cold bath, studying, or exercising before going to bed	No effects reported
Drinking hot water before sleep	Reduced body movements
Taking caffeine	Contradictory effects reported
Fasting	Had calming effects on the first night but disturbed the sleep on the following nights

Figure 12.3 Various factors affect body movements during sleep. For monitoring body movements during sleep, these factors must be controled to reduce the error (adapted from Kleitman 1939).

12.3.3 Some results from monitoring body movements during sleep

Early studies measured the average frequency of movements during sleep. Considering the wide range of factors given on the previous table, we can expect some variations in the reported figures. Choosing the minimal physiological mobility requirement based on those figures would not be optimal. Figure 12.4 gives some average values for the frequency of movements from the studies reviewed by Kleitman (1939).

Subjects	Type of sleep	Average time interval between postural changes
Two to five year old children	Night sleep	7.5 min with long periods of stillness (1 h or more)
56 nursery-school children	79 min nap	25 min (less movements during nap was also reported for adults)
11 adults	Night sleep	11.5 min
College men	Night sleep	12.8 min (range of 7.3 to 21.5)
Middle-aged males	Night sleep	9.0 min (range of 6.3 to 12.5)
Middle-aged females	Night sleep	10.5 min (range of 7.5 to 14.4)

Figure 12.4 Normal people change their postures during sleep quite frequently to relieve pressure (adapted from Kleitman 1939).

Exton-Smith and Sherwin (1961) monitored 50 elderly patients during sleep to study the relation between mobility and pressure sores by counting their major movements using an inertia switch and a counter. Small movements could operate the switch once or twice whereas large movements could operate them up to 12 times. They grouped the patients according to their average number of movements as shown in figure 12.5. These results suggest that using this simple and economical method can be helpful in identifying high risk patients.

Group	No. of patients	No. of movements Group mean (ranges)	No. of patients with pressure sores
0–20	10	10 (4–20)	9
21–50	12	36 (23–50)	1
51–100	19	74 (54–97)	0
101+	9	162 (110–225)	0

Figure 12.5 Elderly patients with small number of movements during sleep developed pressure sores. The direct relation between body movements and pressure sores suggests that a simple counter can be an effective device to detect patients at risk of pressure sores (adapted from Exton-Smith and Sherwin 1961).

Bardsley *et al* (1976) monitored body movements of normal subjects during sleep to evaluate pressure reliefs on three different types of mattresses: foam on spring, foam on board, and bead mattress. They used the technique described above to classify the movements as large and small. Whole body movement was defined as a large movement, others as small movements. They observed that mobility (small and large movements) for each subject increased from foam on springs through foam on boards to the bead mattress. Large movements reduced while small movements increased. They found that mobility scores between subjects were highly variable while there was no significant difference in the number of large movements. They concluded that it should be possible to identify the influence of different supports provided that the sample population is large.

12.4 TRANSFERRING INFORMATION

Most of the wheelchair training devices described in the literature transfer a single bit of information via a buzzer or a source of light such as light-emitting diode (LED). This is basically a warning alarm to remind the patient to perform push-ups at appropriate intervals. These devices usually reset the alarm each time they detect a valid push-up within the interval. Another approach is to measure the pressure–duration distribution on the skin and either reproduce its mirror image on the healthy skin (Rosenberg and Lach 1985) or process the data and generate single-bit prompts to initiate push-ups whenever necessary (Webster 1989). In this section, a single bit of information means a warning alarm and a pattern reproduces the mirror image of the pressure distribution on the skin.

Buzzers and LEDs are inexpensive alarms but are conspicuous and can cause embarrassment for the user (White *et al* 1989). The use of these alarms as a punishment for not performing push-ups also can result in the disabling of the system when there is no inspection by nursing (Klein and Fowler 1981). An ideal device to transfer either single bit or pattern information should: be inconspicuous so it does not cause embarrassment, not act as a punishment, not disturb the daily life, be cost effective, be easy to wear and use, be durable, not need frequent maintenance, be easy to use by different patients with a simple adjustment, be usable with folding wheelchairs, and have convenient connection and disconnection.

Possible inconspicuous alarms for wheelchair training devices are visual displays, electrotactile or vibrotactile stimulators, and effector balloons to reproduce pressure distribution on the healthy skin. Figure 12.6 shows requirements and attachments for each alarm. The following sections describe different methods for transferring information.

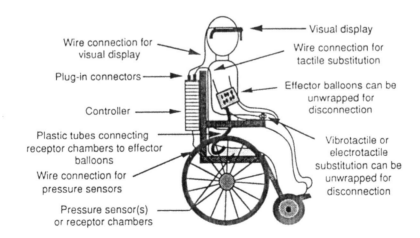

Figure 12.6 Attachments for different types of alarms for a wheelchair training device.

12.4.1 Visual alarms and displays

A single source of light such as a very small LED can transfer on/off information and if mounted properly on the frame or lens of eyeglasses does not attract attention. An LED array mounted on the eyeglass lens can display a pattern. For example, in the Upton eyeglasses for the deaf, mounted lights provide a visual display of speech features related to voiced, fricative, and plosive phonemes (Upton 1968). It is also possible to use a single or an array of LEDs projected on a small reflecting surface on the rear of the lens (Pickett 1974). Visual displays do not require special readjustment from person to person. They have low power consumption and simple ones are inexpensive.

12.4.2 Tactile sensory substitution

Tactile sensory substitution is using tactile sensing to receive information normally received by another sense or at a different area of skin. There are many reports on tactile vision or hearing substitution in which visual or audio signals are converted to digital patterns and transferred to the skin to aid the blind or deaf. Tactile substitution is also used in transferring pressure patterns under the insensate foot of diabetic patients to sensate skin (Wertsch *et al* 1989). The same concept applies in hand sensory substitution in which pressure patterns on the glove or robotic arm are transferred to sensate skin (Bach-y-Rita *et al* 1987). There are different ways to stimulate the skin:

(1) Mechanical vibration
(2) Electric current
(3) Variable pressure (used for transferring pressure–time pattern).

It is simple to use methods (1) and (2) to transfer single bit information. To provide an alarm to the patient, the device should activate the stimulator above the threshold of sensation and below the threshold of pain. Threshold levels vary for different locations on the body and are different from one person to another. Another important factor is adaptation of skin receptors to the stimuli. In general, the skin sensation threshold may increase when stimulated continuously. Thus intermittent stimulation is more noticeable. For transferring pattern information, we should consider spatial resolution, two-point-discrimination threshold, spatial and temporal integration which are summation of the stimulus over some area of the skin and time respectively. Kaczmarek *et al* (1990) review electrotactile and vibrotactile displays for sensory substitution systems and provide 150 references.

Vibrotactile substitution
Vibrotactile transducers can either provide an alarm to the patient to perform push-ups or present pressure–duration information on the healthy skin. These transducers were used for tactile vision substitution in which information from a digitized picture was transferred to a vibrotactile array placed on the abdomen (Kirschner 1986). A vibrating contactor on the skin can produce two types of vibrotactile sensation: flutter and vibration. Receptors found in the upper dermal layer of the skin are responsible for the flutter sensation where the

frequency of the contactor is less than 40 Hz. For frequencies between 80 and 500 Hz, receptors in subcutaneous layers sense the vibration (Kovach 1985). The perceived intensity decreases exponentially with increased time of stimulation and a longer time is required for complete adaptation at higher intensities (Berglund and Berglund 1970). Kovach (1985) determined that using 200-Hz sinusoidal motion for the contactor in a vibrotactile array requires a peak force of 30 mN and a displacement of 200 μm. He compared different vibrators based on their receptibility by the skin, size, weight, energy efficiency, commercial availability, and durability. He compared piezoelectric transducers in different forms (disks, ceramic bimorphs, and film), TO-5 impulse solenoids, and QMB-105 miniature audio transducers. He found that the QMB-105 miniature audio transducers were suitable for tactile arrays.

Kovach (1985) reported some practical problems using a two-dimensional array of audio transducers for vibrotactile vision substitution. The operational frequency was in the audible range (200 Hz) and therefore they were noisy. The array assembly did not allow free ventilation and was uncomfortable. For wheelchair users, this type of transducer would not be a good choice for transferring patterns. For a single-bit alarm, a vibrating transducer can be worn on the wrist like a wristwatch (Webster 1989). The applied frequency should be below the audible range.

Wearable tactile aids are commercially available to assist the deaf to be aware of normal sounds such as doorbells, telephones ringing, and traffic sounds. These tactile aids have vibrators that are worn with wrist harness or on the chest. The vibrator is a tiny loud speaker that produces sensations of feeling based on the input sound. We may easily adapt this type of vibrator for alarming purposes without attracting the attention of others.

Electrical stimulation
A local electric current, injected via electrodes into the skin, can evoke tactile sensation at the location of the electrodes. To inject current, there should be a relatively small electrode to increase the current density above threshold and a return path with a large contact area. Figure 12.7 shows a possible configuration for electrodes. A single or an array of small electrodes (with area of 5–10 mm^2) placed on a common return electrode can inject electric current into the skin to present single bit or pattern information respectively. Depending on the voltage, current, waveform of stimulation, electrode size, and the skin characteristics for each subject, this sensation can be described as touch, itch, tingle, vibration, buzz, pressure, pinch, sharp, and burning pain (Kaczmarek *et al* 1990). The nonhomogeneous nature of the electrode–skin interface often causes high

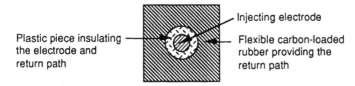

Plastic piece insulating
the electrode and
return path

Injecting electrode

Flexible carbon-loaded
rubber providing the
return path

Figure 12.7 Active center electrode injects electric current into the skin where surrounding dispersive electrode provides the return path. Current density at the location of injecting electrode is relatively high and results in electrical stimulation of the skin. An array of active electrodes placed on a single return electrode can be used for transferring pattern information.

current density and therefore thermal damage to the corneal layer of the skin. Frequent use of electrode pastes can control this pain (Mason and Mackay 1976).

Effector balloons

Rosenberg and Lach (1985) reported their preliminary study on a device which could reproduce the mirror image of pressure–time conditions of the affected area to remote healthy skin. Receptor chambers filled with air or liquid were placed under the weight-bearing skin. The pressure was transferred through tubes to effector balloons placed on the healthy skin. The subject could become aware of a dangerous pressure–time distribution and develop a protective response which is shifting to a different weight-bearing area.

We expect some practical problems for this method which are not addressed by the authors. Wearing a pressing cuff for a long period of time might be inconvenient, uncomfortable, and impractical. We did not find further reports about the effectiveness of this device in reproducing the pressure pattern on the arm, and also how well subjects would sense and react to that.

12.5 STUDY QUESTIONS

12.1 From the experimental graphs of pressure vs time, list the lowest pressure found by each investigator to cause pressure sores.

12.2 Explain the reasons for differences of pressure–duration curves.

12.3 Explain how we could use pressure–duration curves.

12.4 What are the typical procedures for pressure relief training and what is the role of electronic devices?

12.5 Describe an appropriate push-up.

12.6 Specify the actions that a pressure relief training device should perform.

12.7 List factors that affect body movements during sleep.

12.8 Explain how different mattresses affect body movements during sleep.

12.9 Explain the relation between body movements during sleep and pressure sore risk.

12.10 Describe a method to identify elderly patients at risk for developing pressure sores.

12.11   What are single bit and pattern information in pressure relief training devices?

12.12   What are the practical considerations for alarm design?

12.13 Compare different types of alarms with respect to an ideal one described in this chapter.

13

Other Preventive Methods

Annie Foong

In this chapter, we examine some of the less conventional methods used by various investigators in pressure sore prevention. Most of the methods mentioned are not as extensively used or investigated as preventive methods seen in previous chapters. Nevertheless, some of these methods do give promising preliminary results and may establish themselves as useful clinical techniques in the future.

13.1 FUNCTIONAL ELECTRICAL STIMULATION

Between 1986 and 1990, Levine and Kett performed several studies on the effect of functional electrical stimulation (FES) of muscles on pressure sore prevention. Keating reported similar work in 1989 but has yet to report any findings. The studies by Levine and Kett had three objectives:

(1) The effect of FES on seating interface pressure (Levine *et al* 1989b)
(2) The effect of FES on tissue shape variation (Kett *et al* 1986, Levine *et al* 1990a)
(3) The effect of FES on blood flow (Kett *et al* 1988, Levine *et al* 1990b).

A dual channel neuromuscular stimulator provided bilateral stimulation of the gluteus maximus through silicone rubber surface electrodes. The stimulator was controled for current amplitude, pulse rate, and duty cycle. Grimby and Widerstad-Lossing (1989) found that a constant current stimulator is preferred because it provided more consistent and repeatable muscle response. The cathode was placed on the gluteus maximus, and a common anode over the sacrum. Levine and Kett used frequencies from 3 to 50 Hz and varied current amplitude so as to obtain a 25-mm medial-lateral movement of the knee. Various duty cycles were also reported. A commonly used cycle was that of a 4 s on followed by a 4 s off cycle. Levine *et al* (1989b) reported that FES provided tissue undulation and created a dynamic effect for pressure relief. A TIPE pad was used to monitor such pressure variations. Figure 13.1 shows that the stimulation caused more switches to be closed in the TIPE pad at a particular pressure. Since the total load was constant, the increased number of closed switches implied that pressure was effectively redistributed to a larger area.

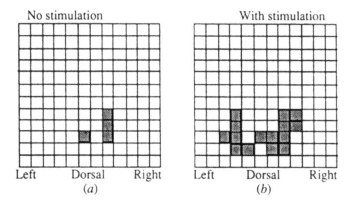

Figure 13.1 With TIPE pad pressure at 50 mm Hg, electrical stimulation closed more switches (*b*) than no stimulation (*a*) (Levine *et al* 1989b).

From their series of experiments, Levine and Kett observed that larger pressure variations was obtained for larger stimulation rates. In addition, their studies with various seating surfaces indicated that pressure changes were maximal with the standard sling, and minimal with the ROHO cushion (most compliant of the given surfaces). Similarly, larger pressure variations were observed for subjects with lesser percentage of buttock adipose tissues. Such an observation is intuitively reasonable. Using the spring-dashpot analogy of figure 5.2 to model tissue or cushion compliancy, compliant materials, with their smaller spring constants are better "shock absorbers". As a result, they respond to FES with smaller amplitude variations.

Levine *et al* (1990b) found that FES also increased blood flow and lymph flow to the muscles, allowing higher oxygen delivery to and metabolite removal from trauma sites. However, they cautioned that increased blood flow to muscles did not imply the same effect in the skin and subcutaneous tissues. They measured blood flow by monitoring the washout of ^{133}Xe and was unable to obtain accurate results for subcutaneous washouts because of Xenon's high affinity for fat.

Finally, ultrasound studies (Kett *et al* 1986, Levine *et al* 1990a) on the shape of buttocks showed that buttock edge contours during FES closely resembled buttock contours suspended in water (figure 13.2). Levine *et al* (1990a) claimed such an observation provided a limited but additional basis of support for the effectiveness of FES as a pressure sore preventive technique.

Besides the immediate noticeable effects of FES, chronic use of FES can increase vascularization, which helps improve the oxygen delivery and metabolite removal process (Salmons 1981). More importantly, FES increases bulk of atrophied muscles and improves muscle tone—which are generally agreed to be factors necessary in reducing the risk of pressure sores.

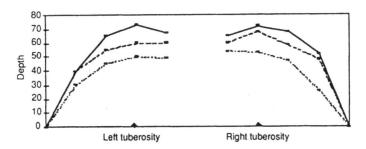

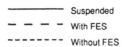

Figure 13.2 Buttock edge contours are flatter when loaded without FES, rounded when suspended, and somewhat rounded when loaded with FES.

However, the negative effects of FES must also be considered. These include increased intramuscular pressure, muscle oxygen requirements, fatigue, heat, and sweat production. Levine *et al* (1990b) reported that oxygen cost per stimulus remained fairly constant as long as stimulation frequencies remained below those required for twitch summation. Fatigue posed a more serious problem. In order that dynamic pressure variation be effective, muscles must be capable of responding throughout the course of a day. Levine *et al* (1989a) found that fatigue was not observed in able-bodied subjects, but was observed for SCI subjects. Fatigue measurements were obtained by noting the number of switch toggles of the TIPE pad. A reduced number of switch transitions would indicate fewer muscle contractions within a given time, thereby indicating that fatigue had set in. Levine *et al* recommended that stimulation should be intermittent, followed by periods of rest. Grimby and Widerstad-Lossing (1989) reported that fatigue was increased for higher-frequency stimulation. Thus, we hypothesize that optimal FES would be that using low frequencies and given in intermittent spurts. Levine *et al* reasoned that the appearance of fatigue in SCI subjects was due to the fact that muscles of these individuals were atrophied and therefore initially unable to sustain the necessary contractions without previous conditioning. Previous conditioning and subsequent chronic use of FES could be expected to increase fatigue resistance (Salmons 1981). Another possible problem is that stimulating the gluteus maximus will cause hip extension and result in the user working out of the chair.

However, figure 13.3 lists other factors not sufficiently discussed by the investigators. We agree with the advantages obtained from chronic use of FES but do question its immediate effects. In particular, we feel that the investigators did not note the difference in effects of FES on muscles which are under pressure and those which are not. Before proceeding further, we include here a summary of physiologic response to FES from an independent source in order to objectively assess Levine and Kett's hypothesis.

Effects of FES	On muscles without pressure	On muscles with pressure
Change of loaded muscle contours shapes to better resemble those of unloaded muscles.	Possible advantage.	Possible advantage.
Chronic use improves muscle bulk and vascularization.	Definite advantage.	Definite advantage.
Circulatory stimulation, increases blood flow.	Definite advantage.	Questionable. Considering capillary pressures to be in the range of 32 mm Hg, blood vessels closed on application of minimal pressures. Therefore, there would not be any blood flow in these vessels, let alone increased blood flow, regardless of whether or not FES is present. Blood flow is probably diverted and increased elsewhere, but not in the regions of high pressures. However, Levine *et al* (1990b) did report that rhythmic contractions of calf muscle enabled a pumping action to drive blood against 90 mm Hg resistance. This may indicate that FES does indeed have an advantage of improving the "pumping" ability of blood vessels.
Increase response for toxic products removal.	Definite advantage.	Questionable. Argument is similar to that of blood vessels. Lymph vessels may be closed under pressure. Removal of toxic wastes is difficult.
Increase demands for oxygen.	Disadvantage.	Disadvantage.
Increase heat and sweat production.	Generally a disadvantage. However, increased heat may also help increase blood flow.	Disadvantage since increased blood flow is still questionable.

Figure 13.3 Critical review of the advantages and disadvantages of FES.

Response to FES includes (Kahn 1987):

(1) Relaxation of spasm.
(2) Monitored contractions of muscles, stimulating active exercise.
(3) Production of endorphins, a body-generated analgesic.
(4) Increased fiber recruitment.
(5) Circulatory stimulation by the "pumping action" of the contracting musculature.
(6) Enhancement of reticuloendothelial response to clear away waste products.

No investigators have yet reported clinical findings that explicitly relate the use of FES to pressure sore prevention. Levine *et al* (1989c) described current clinical trials. However, we must also consider the other advantages of FES, including the ease of use and noninvasiveness of the method. Therefore, if the clinical trials ultimately prove to be successful and remove remaining doubts about the effects of FES, it is very likely that FES will be an effective method in pressure sore prevention.

13.2 SOFT TISSUE AUGMENTATION

Sutton *et al* (1987) developed a procedure to reduce pressure sores over the ischial tuberosities by surgically trimming the bony prominences and subsequently implanting carbon pads over them. Soft tissue padding and other surgical techniques for the purpose of curing pressure sores will be discussed in Chapter 14. The method discussed here is meant for prevention rather than cure. Sutton rationalized that if the load-bearing area of the ischium (or the trimmed remnant) could be made broader and less angular, the pressure in this area would be reduced. The lessened angularity of the bones would also reduce the tissue-shearing effect (see "pinch shear" in figure 2.1(*b*)).

Carbon was used because it is histologically acceptable by human tissues (Jenkins *et al* 1977). In addition, Tayton *et al* (1982) carried out long-term animal studies for 18 months, and confirmed that no malignant changes or carcinogenic effects are found with implanted carbon implants.

Sutton *et al* used pads made of highly structured carbon fibers of about 9 μm in diameter. These were made by carbonizing polyacrylonitrile fibers. Preclinical animal experiments showed that the stress–strain behavior of the pads matched that of the skin within 8 weeks (figure 13.4). In addition, histological examination revealed healthy collagen fibers intertwined with carbon fibers within the implanted pad, and no carbon was found in the para-aortic or lymph nodes. Sutton *et al* reported the insertion of carbon pads in 23 patients. Figure 13.5 shows postoperative pressure recordings which confirm the effectiveness of such pads in distributing ischial pressures over a larger area.

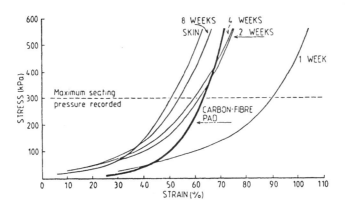

Figure 13.4 Stress-strain curves of implanted carbon pads matched that of skin within 8 weeks (Sutton *et al* 1987).

Postoperative follow-up at an average of 14.8 months revealed that 70% of the patients were functioning well. The others had complications with pads breaking down. Of the seven pads which broke down, five were pads inserted into areas of a previously open sore. There were also cases where the skin over pads broke down. These came about due to careless handling of patient in one case, and the patient not being able to care properly for himself in another. There was also a pad failure due to excessively thickened pads arising from connective tissue infiltration. The implanted pads were removed in all cases. Sutton therefore felt that proper patient selection was important to the success of such a procedure. Considering that surgery for prevention's sake is indeed a drastic measure, we definitely agree that such a procedure should only be performed on the most appropriate patients. Nevertheless, the reported preliminary results were encouraging and do warrant further development of the technique.

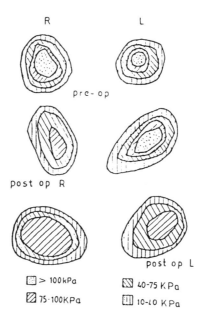

Figure 13.5 Ischiobarograph recordings showed reduced pressures under each ischium postoperatively (Sutton *et al* 1987).

13.3 DRUG-BASED PREVENTION

Irvine *et al* (1961) investigated prior claims that Norethandrolone prevented pressure sores. Norethandrolone has the effect of encouraging weight gain and restoring nitrogen levels. The restoration of nitrogen levels is important to protein formation. Earlier chapters pointed out that weight loss and collagen deficiency are contributory factors to pressure sores. Irvine carried out double blind tests on 395 patients and found no significant difference in the incidence of pressure sores between the patient group treated with Norethanlone and the control group.

Torrance (1983) cited the technique pioneered by A and M Barton in the use of adrenocorticotropin (ACTH) in the prevention of pressure sores in patients undergoing time-consuming surgery.

Barton (1977, 1981) recommended that ACTH be used as a premedication technique for high-risk procedures such as hip surgery or lower limb surgery. They claimed that a single dose of ACTH in gelatin solvent administered 4 h before surgery, prevented cell separation by increasing the adherence of the endothelial cells to one another. When cell separation is prevented, aggregation of platelets and other procoagulant factors to the trauma site is reduced, thereby reducing blood vessel occlusion and tissue necrosis (Chapter 1). This should reduce the chance for pressure sore formation.

Barton and Barton (1981) cited the case of one surgeon, who reported that the incidence of pressure sores was reduced from 25% to zero with the use of ACTH as a preoperative procedure. However, they cautioned that ACTH is a once-only procedure because it liberates some 40 materials from the adrenals—among which are the powerful glucocorticoids—which we have seen in Chapter 1 to have the effect of worsening pressure sores. Glucocorticoids are also known to increase the chance for gastrointestinal ulcers and osteoporosis. In addition, ACTH has immunosuppressive effects, and should not be used for patients with renal failure and hypertension problems.

Pharmacology (Katzung 1989) and physiology (Harper *et al* 1979) texts confirm the fact that ACTH has anti-inflammatory effects, and reduces the formation of eosinophils and lymphocytes at trauma cites. However, we have found no other confirming literature on Barton's recommendation for the prevention of pressure sores. In addition, Frisbie (1986) reported evidence that wound healing of pressure sores is prolonged with the use of anticoagulants. He rationalizes that the accumulation of fibrin at the trauma site is fundamental to tissue repair. The apparent contradiction in evidence reported in literature, together with the many side effects of ACTH, suggest that its use as a preventive measure should be administered with care.

13.4 OTHER TECHNIQUES

Chapter 3 discussed the importance of nutrition as a risk factor in the various scores used for predicting pressure sores. As such, it is reasonable to expect nutrition to be a useful preventive technique. Collins (1983) reported the findings of various investigators who established the relationship between protein and ascorbic acid deficiencies and pressure sore formation. Other dietary requirements, like zinc and iron (Williams *et al* 1988) are important not only to the prevention, but also to the healing of pressure sores (Chapter 14). It is also a common practice at hospitals to put patients on a high-protein diet before surgery to reduce the chance of pressure sores. Mechanic and Perkins (1988) even went so far as to suggest that for patients who are unable or unwilling to maintain adequate nutritional intake, exogenous means of nutritional support should be given. Freyone also cited that Agris (1979) advocated a positive nitrogen balance through a high caloric and vitamin supplementation program. As mentioned earlier, positive nitrogen balance is important to the synthesis of proteins. More specifically, their program aimed at achieving a serum protein of greater than 6 mg/100 ml, and a hemoglobin of greater than 10 g/ml. They did not mention how such high levels of protein and hemoglobin were to be maintained. However, Pinchofsky-Devin and Kaminsky (1986) suggested that oral supplements such as canned polymeric diets be given in addition to a well-balanced meal. They also noted that patients with a serum protein level of less than 3.5 mg/100 ml are considered as mildly deficient; and those with less than 2.5 mg/100 ml as severely deficient in nutrition. The best strategy, as suggested by Pinchofsky, was to have constant nutritional assessment to be performed at least twice a month, and make recommendations for nutritional requirements accordingly. Natow (1983) suggested a high carbohydrate, high protein, moderately low fat diet with adequate calories. In addition, he recommended that sufficient vitamin and mineral intake and daily fluid intake of 800–1000 ml to ensure optimal nutritional and hydration state. We believe that the difficulty in finding a "one-diet-fits-all" program for pressure sore prevention is difficult, if not evasive. This is because patients do not fall neatly into discrete categories of nutritional state. Furthermore, because of differences in metabolism, mobility, state of health, age and other factors, different patients will not have a similar response to any given diet. From the various papers surveyed, the dietician plays an important role in correctly determining the nutritional requirements of patients. Although there was no one standard diet that all investigators agreed on, they agreed to the fact that good nourishment is an important factor in the prevention of pressure sores.

Collins (1983) suggested yet other techniques in which skin care lotions and oils were used for their lubricating and antibacterial properties, and laxatives used for preventing incontinence. Kenedi *et al* (1975) suggested the use of clinically-dried air directed at areas at high-risk for pressure sores. However, no results are reported to confirm their effectiveness.

13.5 STUDY QUESTIONS

13.1 On what reasons was the hypothesis that FES may be effective in preventing pressure sores based?

13.2 Give the various parameters important to a FES device. Suggest values for these parameters. Give reasons for your selection.

13.3 Explain why pressure variations are larger for less compliant seating surfaces.

13.4 Give the negative effects of FES and suggest what might be done to overcome them.

13.5 Why is fatigue observed in SCI patients and not able-bodied subjects? What is bad about fatigue?

13.6 Give the advantages of using carbon for an implantable pad. Suggest other possible materials for such a pad.

13.7 List the pros and cons of using ACTH as a preoperative drug for the prevention of pressure sores.

14

Treatment of Pressure Sores

Todd Jensen

Treating pressure sores is a difficult, time-consuming, and expensive task. Therefore, many research projects seek methods not only to prevent pressure sores, but also to increase the rate and probability of pressure sore healing. This chapter will discuss conventional and alternative methods of treatment that are being studied. Most treatments attempt to control one or more of the factors that affect healing in order to increase the wound healing rate. We first present a description of the physiology of wound healing and the factors that affect wound healing in order to understand why one form or another of treatment is used.

14.1 PHYSIOLOGY OF WOUND HEALING

14.1.1 Four stages

Torrance (1983) outlined the wound healing process in four stages—the inflammatory response, the destructive phase, the proliferative phase, and the maturational phase. Figure 14.1 summarizes the major activities that occur within the wound during each stage. The rate of the healing process and length of each stage are highly variable and depend on many individual factors. The stages do occur in sequence, but there is some overlapping of the activities of each stage (Sieggreen 1987).

Stage	Activities
Inflammatory response	Vasoconstriction initially; blood clotting; vasodilation; arrival of leukocytes.
Destructive phase	Dead tissue and contaminating bacteria removal; new blood vessels grow from wound edges; fibroblasts begin to multiply.
Proliferative phase	Granulation tissue formation; collagen synthesis; epithelialization; wound contraction.
Maturational phase	Diminution of vasculature; enlargement and increase in tensile strength of collagen fibers; wound contraction; scar formation.

Figure 14.1 The major physiological activities during the four stages of wound healing.

Inflammatory response
The inflammatory response is usually marked by the occurrence of redness, swelling, and heat in the injured area. Vasoconstriction occurs within seconds and lasts a few minutes. After this initial response, vasodilation takes place bringing plasma proteins along with other blood constituents into the tissues. This activity peaks at approximately ten minutes (Torrance 1983). Blood platelets aggregate in the wound and trigger blood coagulation to effect hemostasis in the disrupted blood vessels. Hemostasis is limited to the wounded area by several intrinsic factors (Clark 1988). This stage also includes the arrival of leukocytes. Neutrophils are the first leukocyte to arrive and help to rid the site of contaminating bacteria during the inflammatory response. They are short lived and after 2 to 3 days become part of the wound exudate (Sieggreen 1987). Monocytes infiltrate the wound shortly after neutrophils and are converted to macrophages. Neutrophil activity is resolved while macrophages accumulate (Clark 1988). Figure 14.2(*a*) shows the wound shortly after the inflammatory response has begun.

Destructive phase
Debridement (excision of necrotic tissues) is an essential part of the healing process because it provides a sterile field in which cells can migrate and multiply (Barton and Barton 1981). Macrophages ingest and digest pathogenic organisms along with tissue debris and release many biologically active substances that recruit additional inflammatory cells to aid in tissue decontamination and debridement (Clark 1988). The substances released by the macrophages also cause necessary growth factors and other substances to infiltrate the wound to initiate and propagate the granulation tissue that is formed in the next stage (Clark 1988). New blood vessels start to grow from the wound margin and fibroblasts, cells capable of forming collagen, begin to multiply (Torrance 1983). The foundations for cell movement and proliferation have now been formed (Barton and Barton 1981).

Proliferative phase
Granulation tissue consisting of a dense population of macrophages, fibroblasts, and neovasculature embedded in a loose matrix of collagen and other substances appears (Clark 1988). Figure 14.2(*b*) shows the granulation tissue beginning to form. As collagen fibers become more organized, the tensile strength of the wound increases (Torrance 1983). New blood vessels grow (angiogenesis) as in figure 14.2(*c*) and infiltrate the wound simultaneously with the fibroblast and neomatrix components of the granulation tissue (Clark 1988). The new vessels form to nourish the tissues and always originate from existing vessels, beginning as capillary buds (Hunt and Van Winkle 1979). Epithelialization, the migration of cells that cover all the free surfaces of the body, takes place to regenerate the epidermis. Once the epithelial cells begin to migrate, they lose their ability to divide. Therefore, epithelial cells must migrate from areas with undamaged epithelial cells: bases of viable hair follicles, sebaceous glands, sweat glands, or wound margins. The infiltration of new blood vessels and tissues from the wound margin leads to the contraction of the wound (Barton and Barton 1981).

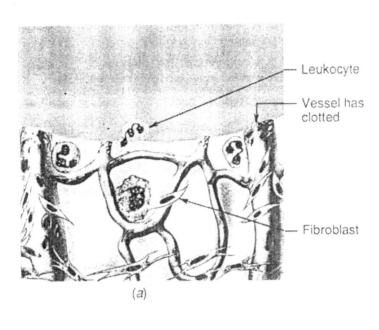

Leukocyte

Vessel has clotted

Fibroblast

(a)

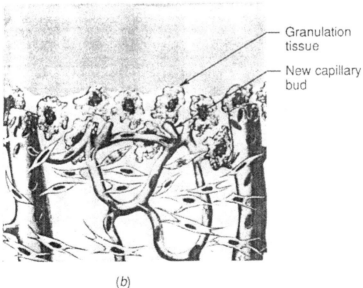

Granulation tissue

New capillary bud

(b)

Figure 14.2 (a) The inflammatory response, which takes place just after injury, is beginning in a full thickness wound (loss of tissue to subcutaneous layer). The blood vessels have clotted and leukocytes leave the vessels and appear in the wound exudate. (b) Granulation tissue forms during the proliferative phase. The predominant leukocytes are macrophagic and stimulate angiogenesis in addition to the fibroblasts and neomatrix substances of the granulation tissue. Capillary buds begin to appear (Hunt and Van Winkle 1979).

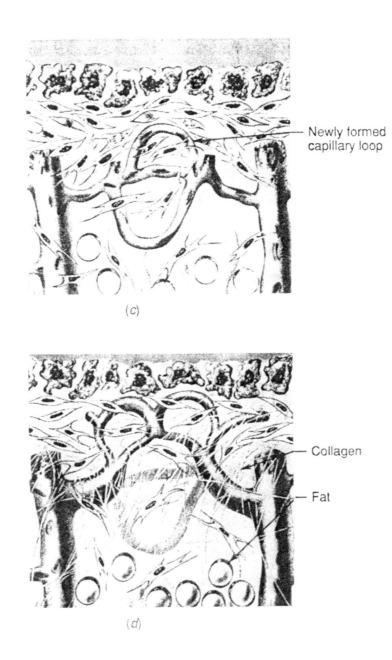

Figure 14.2 (c) During the proliferative phase a new capillary loop forms to supply nutrients to the new tissues. (d) The edge of the wound advances and more new capillary loops form as the supplying artery and vein grow larger (Hunt and Van Winkle 1979).

Maturational phase
Figure 14.2(*d*) shows the wound undergoing remodeling. During this stage the matrix produced simultaneously with granulation tissue is constantly altered following the dissolution of the granulation tissue (Clark 1988). There is a diminution of vasculature (small blood vessels) and an enlargement of collagen fibers. The collagen is reabsorbed as the scar matures, and the scar softens and flattens (Torrance 1983). This phase begins about three weeks after wounding, and can take up to two years or more to complete (Sieggreen 1987).

14.1.2 Factors that affect the rate of healing

Many factors that reduce the rate of healing have been cited. Inadequate nutrition, intercurrent disease, and movement have all been indicated as causes of slower healing rates (Torrance 1983). Hypochlorite solutions, decreases in temperature, high concentrations of sodium chloride, glucose, or sucrose, and large fluctuations in pH levels at the wound site can also be the cause of reduced healing rates. Any treatments should not alter the pH level, or epithelial cells will be affected (Torrance 1983). High levels of sodium or glucose in the wound can inhibit phagocytosis, the process of ingestion and digestion of necrotic tissue and foreign matter by the inflammatory cells (Torrance 1983). Temperature drops as small as 2°C can reduce leukocyte activity to zero (Barton and Barton 1981). Anticoagulants can also delay healing as the end product of the common coagulation pathway, the accumulation of polymerized and stabilized fibrin, appears to be fundamental to tissue growth (Frisbie 1986). Dehydration at the wound site encourages scab formation and destroys the epidermal remnant source of new epithelial cells. Occlusive dressings that keep the wound moist are recommended as long as infection can be controlled (Barton and Barton 1981).

Nutrition
Proper wound healing requires specific nutritional requirements. Proteins are the basic components of all living cells. Protein is broken down into amino acids, and the essential amino acids are not synthesized by the body for metabolic functioning. They must be part of the diet to maintain proper cellular integrity. Wounds that are large, secreting, and/or hemorrhaging may lose up to 100 g of protein per day (Bobel 1987).

Carbohydrates and fats are needed to provide energy to sustain the metabolic processes. Fats provide a more concentrated source of energy by offering 9 kcal/g in comparison to 4 kcal/g for carbohydrates and proteins. Leukocyte activity is promoted by the presence of glucose, the simple form of carbohydrate carried by the blood.

Vitamin C (ascorbic acid) is the vitamin that has the most significant effect on wound healing. Collagen synthesis cannot occur without the presence of vitamin C. It also increases the resistance to infection by promoting leukocyte migration (Bobel 1987). Large doses of vitamin C may be necessary in extraordinary circumstances (Constantian and Jackson 1980a).

Vitamin A is important in wound healing because it enhances epithelialization and granulation (Bobel 1987). It can also overcome the antiinflammatory effects of corticosteroids on epithelialization (Constantian and

Jackson 1980a). Ordinary doses of vitamins B and D are also needed, but vitamin E may have some antiinflammatory effects and probably should be withheld until later stages of healing. A deficiency in thiamine, a B vitamin, may cause a reduction in the capacity of cells to migrate and reproduce during healing (Sieggreen 1987). Vitamin K is an essential blood clotting component and indirectly affects wound healing by preventing hemorrhage and hematoma formation, which can lead to infection and tissue separation (Bobel 1987).

Trace elements such as zinc, magnesium, copper, and calcium also appear to be required for normal cellular function and proper healing. Zinc is the mineral that has the most significant effect. Zinc deficiency leads to a reduction in the rate of epithelialization and cellular proliferation (Bobel 1987).

Oxygen
The lack of oxygen due to injury to the local circulation can prevent the pressure sore from healing. Oxygen is carried by the blood and is needed to produce energy during collagen synthesis and other metabolic processes (Constantian and Jackson 1980a). Smoking cigarettes can reduce the amount of oxygen available to the wound because carbon monoxide from the cigarette smoke has a greater affinity than oxygen for the hemoglobin molecule in the blood. Adipose (fatty) tissue has a poor blood supply, which interferes with the amount of oxygen to the wounds of obese people and leads to a greater incidence of the wound reopening after closing (Sieggreen 1987).

Antiinflammatory steroids and drugs
Any antiinflammatory steroid will suppress repair if administered at or within two or three days of the time of wounding (Hunt 1979). If the desirable inflammatory response is reduced, then subsequent repair of the wound will be slowed. The presence of steroids can also inhibit granulation tissue and fibroblast and capillary proliferation (Sieggreen 1987). Vitamin A and anabolic steroids can restimulate a suppressed inflammatory response, although healing may also be prolonged if too much inflammation occurs because the increasing amounts of inflammatory cells compete with fibroblasts for the limited nutrients at the wound site. Aspirin and other antiinflammatory compounds can also suppress repair (Hunt 1979).

Disease
Diseases can interfere with many biological factors that directly or indirectly affect healing. Cardiac disease can cause poor circulation, which can lead to insufficient delivery of nutrients to the wound site (Torrance 1983). Diabetes also interferes with blood flow and oxygen distribution. Diabetics may also repeatedly injure an area due to impaired sensation. Jaundice (liver failure) and uremia (retention of excessive by-products of protein metabolism) have also been implicated as causes of diminished wound healing (Hunt 1979).

Complications caused by infection
All pressure sores should be considered contaminated wounds, and patients with large wounds may require isolation (Black and Black 1987). An infection may interfere with or stop the progress of healing, and its end result is often to increase the synthesis and deposition of collagen in an undesirable manner

(Hunt 1979). Bryan *et al* (1983) emphasize that pressure sores can be potential sources of many forms of bacteremia in hospital patients and the diagnosis of pressure sores for infection can be difficult. Osteomyelitis (bone disease) associated with a pressure sore may be the cause of infection (Thornhill-Joynes *et al* 1986). Infection in large, deep wounds can cause life-threatening sepsis (the presence of pus-forming or other pathological organisms), and a 50% mortality rate has been reported for such cases in elderly patients (Constantian and Jackson 1980b). Infection of pressure sores can also lead to general sepsis and death in spinal para- and tetraplegics (Guttmann 1976).

14.2 ASSESSMENT AND CLASSIFICATION OF PRESSURE SORES

Pressure sores are classified by severity and physical characteristics during research and treatment in order to assess the progress of healing. Many pressure sores are classified by grades and clinical appearance in the same manner as other wounds. Systems of classification have also been developed primarily for pressure sores.

14.2.1 Types and grades

A simple classification divides pressure sores into partial or full thickness wounds (Fernandez 1987). Partial thickness sores include sores that have destruction of the epidermis and dermis. Full thickness sores involve destruction of the epidermis, dermis, and subcutaneous layer. Barton and Barton (1981) divided pressure sores into three types, Type I, Type II, and Totally Inactive. Type I sores were due to pressure alone, Type II sores were due to local endothelial cell damage in which the necrosis appeared to arise from deep tissues, and Totally Inactive sores were sores that had stopped healing. Figure 14.3 shows the three-part method Lomas (1988) used for the assessment of wounds. First, the wound was given a grade as in figure 14.3(*a*). In the second step the wound was given a color on the basis of its clinical appearance as in figure 14.3(*b*). In the third part of the method, a qualitative measure of the exudate flow was taken as in figure 14.3(*c*).

Guttmann (1976) classified pressure sores by the degree of tissue damage. *Transient* was a disturbance of the circulation in the tissue marked by reversible erythema with some edema. *Definite skin damage* was divided into three cases. The mildest case was marked by the presence of erythema and congestion with discoloration and induration of the skin. Pressure sores that included superficial skin death, exposed cutis vena, and possible blister development were included in the second case of *definite skin damage*. The third and most severe case was marked by necrosis and ulcer formation with possible pigmentation of the border zones of the sore. *Deep penetrating necrosis* included sores that involved the subcutaneous tissues (fascia, muscle, and bone) and may have formed into large, grotesque shapes. *Sinus sores communicating with bursae* included sores that enclosed a sac or envelope lined with synovia and containing fluid. *Closed ischial bursa* was a specific form of pressure sores that were associated with the later stages of paraplegia. They were caused by acute trauma (e.g. bumping buttocks) and were marked by swelling with bloodstained fluid and/or a cavity. The most severe (and rarest) degree of pressure sore was *cancerous degeneration*.

Grade	Physical appearance
1	Discoloration . nd persistent erythema.
2	Superficial skin loss only; blister.
3	Loss of tissue but no cavity.
4	Loss of tissue with a cavity.

(*a*)

Color	Clinical appearance
BLACK	Necrotic area of dead tissue.
GREEN	Infected (pus; offensive odor; inflammation).
YELLOW	Slough-dead tissues accumulated in exudate.
RED	Granulation tissue present.
PINK	Epithelialization with white or pink tissue present.

(*b*)

Measure	Amount of exudate flow
LOW	
MEDIUM	Measured qualitatively.
HIGH	

(*c*)

Figure 14.3 The three-part method of assessment of pressure sores used by Lomas (1988). The wound is given (*a*) a grade according to the physical appearance, (*b*) a color according to the clinical appearance, and (*c*) a measure based on the qualitative measurement of exudate flow.

14.2.2 Measurements

Measurements of the physical dimensions of pressure sores are taken to record the rate of healing in a quantitative manner. Area, volume, and temperature gradients are some physical parameters that can be measured. Barron *et al* (1985) made cross-sectional measurements and recorded the shape of pressure sores on a scaled grid. From this they calculated the surface area. Akers and Gabrielson (1984) projected color slides of pressure sores taken at a standard distance onto a summagraphic bit pad. The pad was interfaced to a microcomputer for area analysis. Boykin and Winland-Brown (1986) made direct measurements of diameter and depth of pressure sores using a wound size measurement scale. Barton and Barton (1981) describe how radiometry and thermography can provide information on wound contraction. Allman (1989) suggests taking color photographs to record the progress of healing over time.

The rate of healing is usually measured as the time it takes the wound to completely heal. Stefanovska *et al* (1989) found the wound area decreases exponentially over time as in figure 14.4 and made measurements of the area and depth of the wound over time to find a time constant of healing. They described the healing process using the following equation:

$$S = S_0 \, e^{-t/\tau}$$

where S_0 is the initial area of the pressure sore (cm^2), S is the area at time t, and τ is the time constant of healing (weeks). Positive time constants indicate the pressure sore is becoming smaller, while negative time constants indicate the pressure sore is growing larger. The time constant τ depended on the depth and surface area of the pressure sore. A time constant dependent on the area of the pressure sore, τ_a, and a time constant dependent on the depth, τ_d, were found by plotting the surface area and depth on time scales and then using the following equations:

$$\tau_a = \frac{t}{\ln(S_0/S)}$$

$$\tau_d = \frac{t}{\ln(d_0/d)}$$

where d_0 is the initial depth (mm), and d is the depth at time t. The overall time constant of healing was then calculated as

$$\tau = \tau_a + \tau_d.$$

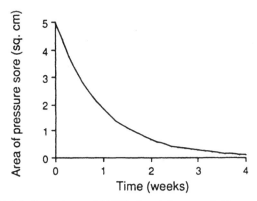

Figure 14.4 Stefanovska *et al* (1989) plotted the area of skin a pressure sore covered over time and found the area decreased exponentially over time. The decrease in area was most rapid initially and then slowed over time.

14.3 CONVENTIONAL TREATMENTS

In any case, the preferred method of treating pressure sores is prevention because current methods of treatment are frequently time consuming and expensive, and they may not reduce the risk of recurrence if a patient remains immobile (Allman 1989). Hospital staffs should know and use methods to help prevent pressure sores (Kostunik and Fernie 1985). McDougall (1976) states that many times patients that had never developed a pressure sore at home developed a pressure sore a few days after being admitted to the hospital. They were afraid to be as mobile as they were at home (e.g. getting a book, reaching a nearby toilet) for fear of being reprimanded by the supervising nurse or upsetting a neatly made bed. Therefore, it is important that the staff be well versed in the problem of pressure sores.

Another important means of prevention is patient education, which should include guidelines that address what patients can do to eliminate, diminish, or counteract the contributing factors that lead to pressure sores (Boykin and Winland-Brown 1986). Early recognition of areas of potential sores and daily skin inspections are two techniques used by patients to prevent the development of pressure sores (Harries 1987). Other methods of prevention and pressure relief have been discussed in previous chapters.

Before beginning the treatment of a pressure sore, many factors need to be considered. Constantian and Jackson (1980a) list eight primary considerations that should govern therapy:

(1) Is the wound thickness partial or full?
(2) What is the degree of contamination or infection and is there a presence of cellulitis?
(3) Is an abnormal underlying bursa present?
(4) What is the extent of undermining of the surrounding tissues?
(5) What is the condition of the surrounding skin?
(6) Are abnormal communications with the pressure sore present?
(7) What is the local bone anatomy and is a heterotopic bone present?
(8) How great is the willingness or ability of the patient to cooperate with the treatment plan?

The answers to these questions help to choose the proper course of treatment. For example, any infection present needs to be controlled in order to allow the pressure sore to heal properly.

14.3.1 Three steps of management

When a pressure sore develops, it must be cared for. Torrance (1983) divided the management of pressure sores into three steps—the removal of pressure, the treatment of any predisposing factors, and the care of the wound. Figure 14.5 presents a flow chart of the process used to treat a pressure sore. Each of the steps will be described in greater detail.

Remove pressure
To prevent further complications and allow the healing process to begin, pressure must be removed from the sore to permit local tissue perfusion and oxygenation to occur (Seiler and Stähelin 1985). Previous chapters described many devices designed for this purpose. A problem occurs because none of the currently available devices can relieve pressure from a contact area with 100% effectiveness (Mulder and LaPan 1988). Good turning protocols can also be used by hospital staff to relieve pressure (Boykin and Winland-Brown 1986).

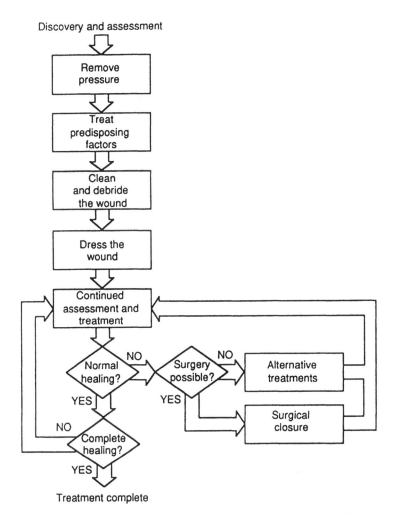

Discovery and assessment

Figure 14.5 The treatment of pressure sores presented as a flow chart. After the pressure sore is discovered, any pressure applied to the area of injury must be removed to prevent further insults. Any predisposing factors that can delay healing must be treated. The wound is assessed and then must be cleaned and dressed in an appropriate manner. Surgery may be needed to close and speed the healing of large pressure sores. If one course of treatment does not work, alternative treatments may be attempted.

Treat predisposing factors

Factors that can delay healing such as those discussed in Section 14.1.2 must be addressed and accounted for. The general health, mobility, and nutritional status of a patient are major factors contributing to normal or delayed wound healing. Chronic illnesses should be treated to prevent any further delays in healing (Barton and Barton 1981). Patients with pressure sores often have problems with wound healing due to inadequate nutrition, and nutritional deficiencies should be assessed and treated. Anthropometric and biochemical measurements along with the physical signs suggestive of nutritional deficits can be used to target particular nutritional losses (Bobel 1987).

A careful examination of the cardiovascular system should also be made as it is necessary that an adequate supply of blood reaches the healing tissues (Barton and Barton 1981). If the patient is anemic, transfusions should be given (McDougall 1976). Any problems with the peripheral vasculature that supplies the wound should be assessed and corrected. Rings should be dispensed of as they act as a tourniquet to obstruct the arterial blood supply and cause edema, which should be prevented because it prevents metabolites from being exchanged (Barton and Barton 1981).

Wound care
Preparation of the wound is the first stage of wound care. A dressing cannot be applied without first cleaning the wound and removing necrotic tissue or ineffective and even harmful results such as infection can occur (Mulder and LaPan 1988). Necrotic tissue does not allow epithelialization to take place and may obscure an infection, which could penetrate into deeper skin layers and cause further complications such as osteomyelitis or sepsis (Seiler and Stähelin 1985). Many cleansing agents are available and include normal saline 0.9%, Savlodil (chlorhexidine gluconate 0.015%, centrimide 0.15%), Eusol (chlorinated lime 1.2%, boric acid 1.25% in water), and hydrogen peroxide. Murray (1988) warns that care must be taken when selecting a cleansing agent as many of them can have detrimental effects on healing. Savlodil appears to be toxic to fibroblasts, key cells responsible for laying down the collagen-based scar in soft tissue repair. The debriding action of Eusol is not specific to necrotic or sloughy tissue, and the use of Eusol can lead to increased urea and acute oliguric renal failure. Hydrogen peroxide may be caustic to the surrounding skin. Fowler (1987) suggests using mild skin-wound cleansers such as nonirritating liquid preparations to assist in the removal of foreign material without delaying healing.

Surgical debridement may be necessary to transform a closed, infected wound to an open, clean wound (Guttmann 1976). Another common procedure of debridement is to apply a wet-to-dry dressing. A saturated dressing is loosely packed into the wound and when dry it is removed, pulling the loose necrotic material with it. Fluids may be forced into the wound using a water pik to dislodge loose debris (Fowler 1987).

All wound care should be done using sterile techniques to prevent the development of septicemia, and antibiotics should be prescribed when quantitative cultures of wound tissue biopsies have bacteria greater than 10^5 organisms per gram of tissue as this level is considered diagnostic of invasive infection (Black and Black 1987). Appropriate topical therapy should be satisfactory in controling infections (Sugarman 1985). Potent but nonirritating local disinfectants can be used to control infections (Seiler and Stähelin 1985). The topical agent chosen as the local disinfectant must not cause pain or impair granulation, nor should it change the appearance of the healing tissues. The appearance of granulation tissue signals when the use of local disinfectants should be stopped, as they can damage granulation tissue.

For most cases, it is recommended that an occlusive dressing be applied to the clean wound as the primary dressing because occlusive dressings provide an optimized wound environment and a means of protecting the wound from outside contamination (Mulder and LaPan 1988). Occlusive dressings create a moist environment which allows epithelial cells to migrate and should be gas permeable to allow the healing tissue an adequate oxygen supply (Seiler and Stähelin 1985). If the pressure sore shows clinical signs of infection, has a culture that yields greater than 10^5 organisms per gram of tissue, has exposed tendons or bones, or has draining sinus tracts, then an occlusive dressing should not be used (Mulder and LaPan 1988). It is also better to use a dry dressing if the sore is inactive (Torrance 1983).

Many synthetic occlusive dressings are available and can be divided into four groupings, each with its advantages and disadvantages (Mulder and LaPan 1988). Figure 14.6 points out the advantages and disadvantages of each type of synthetic occlusive dressing when used on pressure sores. Hydrocolloid dressings are adhesive, gel-producing, water-impermeable membranes. They can be left on for up to one week, and commonly have an adhesive wound contact face, an impermeable outer face, and an exudate-absorbing component that is usually carboxymethylcellulose. They are easy to apply, but may be messy and disruptive when used on high-exudate wounds. Polyurethane and thin film dressings are transparent-adhesive dressings that are often difficult to use. They can be troublesome to apply and may be disrupted or removed by patient movement. It is also common to have exudate leakage due to the minimal absorption of the dressing. Bio-dressings and gels are similar groupings of dressings. They are usually hydrogels of water and polyethylene oxide, have a reinforced polyethylene film, and have a high water content. They are more appropriate for use on abrasions and superficial wounds than pressure sores.

Dressing	Advantages	Disadvantages
Hydrocolloid (e.g. Duoderm)	Easy to apply; good retention of moderate exudate; water impermeable; self-adhesive.	Leakage and displacement with high-exudate wounds.
Polyurethane (e.g. Tegaderm)	Good on superficial wounds.	May be difficult to apply; poor adherence; minimal absorption.
Biodressings (e.g. Spenco Second Skin)	May be used on friable tissue.	Suitable for superficial lesions only; difficult to maintain over wound.
Gels	Same as biodressings.	Same as biodressings.

Figure 14.6 The advantages and disadvantages of the four types of synthetic occlusive dressings (Mulder and LaPan 1988).

Murray (1988) found that nonadherent dressings consisting of thin fiber mixes packed between one or two thin layers of perforated polymeric film were the most popular type of dressing for pressure sores. These dressings are designed for low-exudate wounds as moderate or high exudate can permeate the dressing and allow bacteria to penetrate the wound. This can cause dehydration of the wound and the dressing may adhere to the wound. Murray also found that patients being treated at home may not be getting appropriate dressings due to restrictions on prescribing certain dressings for home use. Murray suggested that many times nurses have a bias towards certain products due to some previous success as many of the dressings used do not possess the characteristics needed for use on a wound. Figure 14.7 describes the criteria that a wound dressing should have to achieve an optimal wound environment and healing rate.

Properties of an Ideal Wound Dressing
• Maintains high humidity between wound and dressing.
• Removes excess exudate and toxic compounds.
• Allows gaseous exchange.
• Provides thermal isolation for the wound surface.
• Is impermeable to bacteria.
• Is free from particles and toxic wound contaminants.
• Allows removal without causing trauma when the dressing is changed.
• Prevents tissue desiccation (rupture).
• Facilitates rapid cell migration.
• Protects newly formed and forming granulation tissue.
• Protects from abrasion.
• Controls odors.
• Can be used as a packing and/or a vehicle for medication.
• Is easy to use (application and removal).
• Is cost and time effective.
• Stays in place.

Figure 14.7 The properties that an ideal wound dressing should possess to create an optimal wound healing environment according to Turner (1982) and Fowler (1987).

To ensure that granulation tissue growing from the base fills the wound space and to prevent the wound edges from healing across too quickly, packing is sometimes used, although it has many disadvantages (Murray 1988). The material used (e.g. ribbon gauze) may be painful to insert and remove and must be inserted loose enough to allow free drainage from the wound.

Lomas (1988) described the use of a wound care flow chart for the care of pressure sores. After a pressure sore was graded and coded by color as described in figure 14.5, an appropriate product and implementation was selected according to the grade, color, and flow of exudate from the pressure sore. The chart was supplemented with handouts that instructed the staff on how to use the appropriate products.

14.3.2 Surgery

Some large pressure sores are treated using surgical methods. Surgery is an alternative to the long process of healing an open pressure sore must undergo, and any full-thickness sore that is greater than 2 cm in diameter should be closed surgically if the patient can tolerate surgery (Black and Black 1987). The restoration of a supple cushion of well-nourished soft tissue and the restoration of sensitivity in supporting regions are two goals that surgery should meet (Reichert 1986).

The wound should be clean and somewhat reduced in size before surgery can take place, and this preoperative period can last two to three weeks (Buntine and Johnstone 1988). Pers *et al* (1986) suggest using a more detailed than usual antibiotic treatment based on bacteria sensitivity to avoid postsurgical complications such as bleeding and infection. Surgery is a two-stage procedure.

The first stage is a complete excision of the pressure ulcer, and the second stage is closure, which can be done at a later time with good results.

Three methods of closure are used. Direct closure is a method that reconnects the skin and deeper tissues. This method leaves a scar over a weight-bearing zone. Buntine and Johnstone (1988) found direct closure does well in most instances, but Black and Black (1987) note that this procedure is prone to reopening because the tension placed on the wound is frequently too great. Figure 14.8 diagrams the second method, in which large random cutaneous (skin) flaps are used to place the scar away from pressure areas. Nearby tissue is elevated from the fascia using a technique called undermining and rotated to cover the pressure sore. The flaps are supplied with nutrients mainly by subdermal vessels, and this tenuous blood supply makes healing unpredictable (Black and Black 1987).

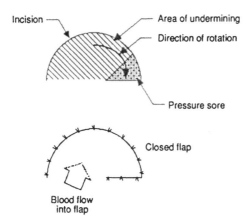

Figure 14.8 The elevated (undermined) skin and subcutaneous fat are rotated over the pressure sore. Nutrients are supplied by blood that enters the reattached area carried by blood vessels that run through the section of skin remaining attached to the original site (Black and Black 1987).

The third method uses myocutaneous (muscle and skin) flaps to obliterate a cavity and increase the blood supply. Figure 14.9 shows an schematic drawing of how a muscle flap is used to surgically close a sacral ulcer. Muscle or portions of muscle and skin can be moved into the wound to close it. This is beneficial because it replaces lost tissues, creates padding, and reduces the risk of new pressure sores (Black and Black 1987). Kauer and Sonsino (1986) stress that a minimal amount of skin and muscle should be used during transposition techniques in case of possible recurrences. Rosen et al (1986) note that the use of muscle flaps is better than large, regional skin flaps because muscle flaps are more reliable, heal more rapidly, and cause fewer complications. Infected parts of bones may have to be excised in certain cases (Guttmann 1976). Any removal or recontouring of underlying bony prominences to reduce pressure should be done conservatively (Rosen et al 1986). If a free flap, one totally disconnected from its original site, is used, care must be taken to reconstruct the arteries, veins, and nerves. This is complex, and therefore, free flaps are only used when simpler flaps cannot be used (Black and Black 1987).

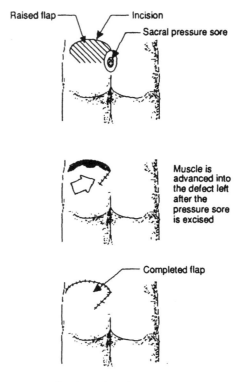

Raised flap —   — Incision
                       — Sacral pressure sore

Muscle is
advanced into
the defect left
after the
pressure sore
is excised

— Completed flap

Figure 14.9 A myocutaneous flap is used to close a sacral pressure sore. The sore is first excised, and then the planned flap is moved into the defect and closed (Constantian and Jackson 1980c).

Black and Black (1987) pointed out that diligent care is needed to ensure that the flap heals properly and described some of the causes of flap failures. Small arteries feeding the flap are stretched and may be occluded by tension on the flap. Blood supply may also be cut off by hematomas beneath the flap that distend the tissues or by pressure from dressings, braces, or bed. Venous stasis promoted by gravity does not provide the tissues with adequate oxygen. Necrosis may occur in flaps with arteriosclerotic veins.

14.4 ALTERNATIVE AND POSSIBLE FUTURE TREATMENTS

Throughout history, there has been a need for treatments to cure pressure sores and other wounds. Ancient Egyptians bound ox fat to wounds, and Indian physicians (vaidya) of the 4th century B.C. auscultated wounds to detect the vayu (inner wind) that caused the wound and would indicate its prognosis (Constantian 1980). A 15th century recipe for wound care mixed March barley and one-half bushel of toads. This mixture was boiled together and fed to a hen that had newly hatched chicks. The patient then ate the hen (Constantian 1980).

In search of better forms of treatment for more complete and faster healing, many new and sometimes seemingly unorthodox methods have been developed. Methods for which faster healing rates have been found include applying some missing factor to the wound such as sugar in a topical form, electrical stimulation, ultrasound, and heat lamps. Treatments that use collagen injections, drug delivery systems, or growth factors are possible future methods that are being explored. Increases in the rate or quality of healing have been reported when using alternative treatments, but the results may be misleading due to problems in the studies used to compare new to conventional treatments. We will discuss these problems after describing some of the alternative and future treatment methods.

14.4.1 Topical applications

Many researchers hypothesize that some factor is missing from present treatment methods, and an increase in healing rates should occur when the proper agent is added (Constantian and Jackson 1980a). One of the agents tried is aloe vera. It has been applied to wounds, and reports of its beneficial effects date back to Hippocrates and Alexander the Great (Cuzzell 1986). Recently, it has been shown that chemicals in aloe vera produce an anesthetic reaction, kill bacteria, and increase local microcirculation in the tissue.

Granulated sugar is another agent that has been applied to pressure sores with the rationale that the pressure sores lack proper nutrition (Constantian and Jackson 1980a). There are many hypotheses as to why sugar therapy has worked (Torrance 1983):

(1) Sugar provides an ideal medium for bacteria growth and spares the tissue
(2) Bacteria gorge on the sugar and then die
(3) The pH of the wound is changed in a beneficial way
(4) Serum and nutrients are brought to the wound by reversed lymphatic flow
(5) Sugar has a stimulating effect on granulation.

A mixture containing Crisco (vegetable shortening), sugar, and povidine-iodine was used on clean wounds with an epidermal element present with some success (Cuzzell 1986). Reasons given for its effectiveness were the shortening sealed the wound and promoted a moist surface, the sugar created a hypertonic environment which was slightly antibacterial, and the providine-iodine was even more strongly antibacterial.

Constantian and Jackson (1980a) list many agents and the hypotheses for their use to promote the healing of pressure sores. Honey has been applied as a cleansing and healing agent. Gelatin foam has been used to provide a presumedly missing matrix for the ingrowth of new cells. Topical porcine collagen was applied to provide a latticework for granulation tissue growth. Topical insulin was used on the basis that it would stimulate protein synthesis. Dried blood plasma and balsam paste was applied with the rationale that it would provide deficient nutrients. Silvetti (1981) applied topically a balanced solution of nutrients (salts, amino acids, polysaccharide, and ascorbic acid) after daily debridement to achieve healing of severe, long-enduring wounds. It was assumed that certain nutrients not available in quality or quantity from the local circulation are needed to repair tissue.

14.4.2 Thermal therapies

Three direct physiological effects heat has on tissue are increased local and remote blood flow due to temperature rise or fall, stimulation of the neural receptors in the tissues, and an increase in metabolic activity. The extent of the physiological response is determined by many factors including the size of the area exposed, the intensity of radiation, the relative depths of absorption of specific radiations, the amount of temperature variation, the rate of temperature change, the duration of the temperature elevation or reduction, and the health and age of the patient (Wadsworth and Chanmugam 1983). The dosage of heat is dependent on the intensity, duration, and frequency of application. Methods of treatment such as ultrasound and electrical stimulation that may produce heat in addition to other effects have been used to treat pressure sores and will be discussed later, but the application of heat only through such methods as hydrocollator packs, hot compresses, paraffin wax, hot water baths, and electric heating pads is not usually recommended for use on infected or open wounds as heat will increase the infective activity (Wadsworth and Chanmugam 1983).

Sometimes a heat lamp is used to dry the wound (Cuzzell 1986). This method should be used with caution as fragile cells in deep granulating wounds may be damaged and skin that is not involved with the pressure sore may be traumatized. It is recommended that a heat lamp only be used on clean, shallow wounds with some sign of epithelialization and on patients who have unaltered sensation and can complain of discomfort.

It has been found that pressure sores heal slowly in cool conditions (Murray 1988). Leukocyte activity can cease with a temperature drop of 2°C from body core temperature (37°C) and can take up to 4 h to peak after the temperature rises back to its previous level. Therefore, it is recommended that dressings should not be removed in advance of inspections to avoid any unnecessary heat loss, and frequent cleanings with cold solutions should also be avoided.

Although it has been stated that pressure sores heal slowly in cool conditions, some researchers have found that cold therapy (cryotherapy) produces beneficial results. Marshall (1971) treated children with pressure sores by holding an ice pack on the sore for up to ten minutes a day, seven days a week. Some sores for which other treatment methods failed healed completely.

14.4.3 Electrical stimulation

Gold sheets have been applied on wounds to speed healing rates, and it is thought that the beneficial effects were from an electrostatic or electrochemical influence established by the interface of the wound and the gold (Akers and Gabrielson 1984). Jercinovic *et al* (1989) found a higher potential (28–41 mV) exists between areas of intact and injured skin than between two areas of intact skin. Electricity has been used as a form of treatment to increase healing rates by supposedly supplementing the natural, endogenous currents taking place during the healing process, although it is not yet known exactly how electrical stimulation increases the rate of tissue regeneration (Vodovnik and Stefanovska 1989).

Most forms of electrical stimulation presently use an active electrode with a positive or negative polarity relative to the body and a ground electrode to cause current to flow across the pressure sore. Electrons carry the charge in electrical wires, but ions carry charge in the body. Electrodes are used to transfer current from electrical wires to the body, and ideally they should be easy to apply, conform to the body, have minimal impedance, have a uniform current density, and maintain these characteristics for many days without irritating the skin (Patterson 1983). Skin provides the majority of the total impedance seen by the stimulator. The resistivity of muscle tissue, blood, and body fluids do not contribute much to this total impedance (Patterson 1983).

The applied current is sometimes referred to as being galvanic or faradic (Patterson 1983). Galvanic current usually refers to a direct current (DC), which may be produced by a battery, and produces a net charge transfer. With faradic current there is no net charge transfer at the electrodes. Faradic current is an alternating current (AC) coupled output signal in which the same amount of charge flows alternately in both directions. The magnitudes of the current in each direction may be different, so long as the lower current is applied for a longer time so that the charge transferred in each direction is equal. The charge transferred may be found using

$$Q = It$$

where Q is the total charge (C), I is the current (A), and t is the time (s).

The charge delivered may be controlled by varying the magnitude of the current used and the application time during electrical stimulation treatments. The power dissipated in the skin may be calculated by

$$P = VI$$

where P is the power (W), V is the potential across the skin (V), and I is the current (A). Substituting Ohm's law ($V = IR$), the above equation becomes

$$P = I^2R$$

where R is the resistance of the skin (Ω).

Wounds have reportedly been treated successfully using low intensity DC currents. Carley and Wainapel (1985) applied a DC current between 200 and 800 μA for periods of 2 h, twice a day, five days a week. During the first three days the negative polarity active electrode was put on the wound site and the ground electrode was attached 15 to 25 cm proximal. After three days and until the wound healed or reached a healing plateau, the active electrode was of positive polarity. The patients undergoing this method of electrical stimulation treatment had faster healing rates, stronger scar tissue, less colonization in highly contaminated wounds, and no further debridement.

Alvarez et al (1983) used a DC current that decreased linearly from 300 to 50 μA during a 24-h treatment period to treat superficial wounds on pigs. They found a significant increase in healing and also an increase in protein and collagen content and an increase in the proliferation (DNA synthesis) rate. Barron et al (1985) used a method they called Micro-Electro Medical Stimulation (MEMS) to administer low-intensity currents across the sore through probes placed 2 cm from the edge of the sore. Possible reasons for the significantly increased healing rates resulting from using MEMS included increased generation of intracellular ATP (the highest energy producing substance for the cell) and accelerated conversion of amino acids to protein.

High-voltage galvanic stimulation produces a current with a high peak intensity during a very short pulse duration so the average current is low (Wadsworth and Chanmugam 1983). It has also been used to obtain significantly higher healing rates (Akers and Gabrielson 1984, Alon *et al* 1986, Feeder and Kloth 1985). A pulsating high voltage is applied to create a pulsating DC current to flow through the pressure sore. The voltage is adjusted to a level just below that which causes visible muscle contractions. Brown and Gogia (1986) suggest that this method is an effective modality for wound healing during acute phases if the polarity of the voltage is negative, and it is inappropriate for subacute phases. Hypotheses for the increased healing rates resulting from use of high-voltage galvanic stimulation include the possibilities that external positive current raises a positive potential that exists in the pressure sore to hasten the repair progress, stimulation with the negative pole destroys bacteria, and high-voltage current increases superficial circulation (Wadsworth and Chanmugam 1983).

Kambic *et al* (1990) stimulated pressure sores in denervated limbs of Hanford mini-pigs. They applied 30–80 μA/cm^2 dc current density 2 h daily for 28 days. Young's modulus in the direction parallel to current flow was lower than that perpendicular to current flow. Although stimulated skin healed faster, the scar did not achieve the overall tensile strength of the normal skin.

Transcutaneous electrical nerve stimulation (TENS) has been used to successfully treat pressure sores (Kaada 1983). A positive electrode was placed between two toes on each foot, and a negative electrode was placed between two fingers on each hand. The intensity of the stimulus was increased until muscles adjacent to the electrodes contracted without pain. Favorable results were obtained for wounds located on various areas of the body, and he suggested that the primary cause of healing was most likely the increased microcirculation resulting from stimulation.

14.4.4 Electromagnetic energy

Electromagnetic (EM) energy in the form of infrared radiation has been indicated for slow-healing indolent wounds because it can increase vasodilation by its thermal effects (Wadsworth and Chanmugam 1983). Infrared radiation has a wavelength between 770 and 15,000 nm. The infrared rays are generated using luminous or nonluminous generators. Luminous generators are incandescent lamps with tungsten or carbon filaments. The filament is placed in a glass bulb which contains an inert gas or a vacuum. Rays with wavelengths from 392 to 800 nm are emitted as current passes through the filament and produces heat. Nonluminous infrared generators usually consist of a wire wrapped around a cylinder of insulating material. The infrared rays produced by either method are directed by parabola-shaped lamps to create a floodlight or spotlight beam. Infrared radiation is not recommended for infected wounds because the increase in temperature can raise the infective activity.

Ultraviolet rays, a form of EM energy produced by the sun, mercury vapor lamps and some fluorescent tubes, are also used to treat some wounds

(Wadsworth and Chanmugam 1983). During treatment the lamps used can heat up to several hundred degrees Celsius and must be cooled. Air-cooled or water-cooled lamps are usually used to treat pressure sores because they are designed for local area applications. Figure 14.10 lists the levels of dosages of ultraviolet light. The dosages are graded into four levels of erythemal reaction observed on normal skin according to the intensity of the reaction. The intensity of each successive dose must be progressed (raised) to cause the same reaction because normal skin thickens in response to the application of ultraviolet rays. Dangers from using ultraviolet rays include eye damage, burns due to overdosage, and sensitizing drugs.

Dose level	Length of ultraviolet exposure required to produce mild erythema (h)	Time for erythema reaction to disappear (days)
E_1 (1st degree)	6–8	1
E_2 (2nd degree)	4–6	2
E_3 (3rd degree)	2–4	3–4
E_4 (4th degree)	2–4	≥ 7

Figure 14.10 Dosages of ultraviolet rays based on the erythemal reaction observed on normal skin (Wadsworth and Chanmugam 1983).

To keep incipient pressure areas from breaking down and producing an open sore, ultraviolet rays may be applied to improve the skin condition, stimulate growth of epithelial cells, and destroy surface bacteria. An E_1 dose, progressed daily, or an E_2 dose for thicker skin in areas such as the heels and ankles is recommended. Granulation tissue growth, epithelialization, and infection prevention in open wounds can be improved using an E_1 dose. The granulation tissue should receive unprogressed E_1 to prevent overgranulation, but the surrounding skin is generally treated with daily progressed E_1. Infected wounds may be treated as uninfected wounds, although higher levels of dosages may be used to remove the slough (infected material).

Pulsed short-wave therapy is a form of EM energy that operates using frequencies centered at 27.12 MHz. The narrow band is an internationally recognized range of frequencies for equipment designed to radiate EM waves for purposes other than communication (Oliver 1984). Figure 14.11 shows a typical drum electrode that is commonly used to generated the pulsed short-waves. The high-frequency current through the coil produces an electric (E) field and magnetic (H) field. The current is pulsed on and off to prevent thermal effects. Some tissue heating occurs during the on phases, but the off phases are of long enough duration to allow complete cooling (Hayne 1984). E fields create H fields within the body tissues, and similarly an H field creates an E field (Hayne 1984). Some of the effects created by pulsed short-wave therapy that stimulate wound healing include the increase in the number of white cells and fibroblasts, absorption of hematoma, reduction of inflammation, increase in the rate of fiber orientation and collagen deposition, and earlier collagen layering.

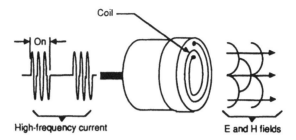

Figure 14.11 A typical drum electrode used to create short-wave EM energy. Electric (*E*) and magnetic (*H*) fields are created when high-frequency current flows through the coil (Oliver 1984).

14.4.5 Oxygen and air therapy

Circulating air has also been used to treat pressure sores. Isherwood (1976) noted increased healing rates when air was filtered to remove bacteria and continuously supplied over pressure sores at rates of 25 ml/min to 2 l/min. Keeping ambient air out from the area of the pressure sore is very important to prevent contamination.

Hair dryers have also been used on the basis that they would promote increased circulation (Constantian and Jackson 1980a). Pressure sores have also been exposed to negative air ions with the rationale that the ions would encourage granulation tissue and epithelial growth (Constantian and Jackson 1980a). Hyperbaric oxygen has been used with the rationale that pressure sores are hypoxic (Constantian and Jackson 1980a). High-pressure, pure, humidified oxygen is applied by a chamber which forces 100% oxygen into the superficial tissue (Nawoczenski 1987). Studies have shown healing rates may be increased and surgical flaps resuscitated by administering hyperbaric oxygen (Eltorai 1981).

14.4.6 Ultrasound

Ultrasound is a form of acoustic vibration at frequencies greater than the hearing range of humans (20 Hz – 20 kHz). Ultrasound applied in medical applications is usually of a frequency between 1 and 15 MHz and is produced by the piezoelectric effect (Wadsworth and Chanmugam 1983). Figure 14.12 shows a typical ultrasonic transducer and generator. The generator produces an AC current at a frequency equal to the resonant frequency of a quartz crystal. At this frequency the crystal vibrates with an amplitude that depends on the thickness of the crystal and intensity of the current. A transducer head placed in front of the crystal vibrates mechanically due to the acoustic vibration of the crystal.

Ultrasound has mechanical, thermal, chemical, and electrical effects (Wadsworth and Chanmugam 1983). Some of the mechanical effects (micromassage) include loosening of the microscopic cell structure, friction, oscillation of particles in a liquid, intracellular massage, and the transport of drugs. Vasodilation and acceleration of lymphatic flow are the main thermal effects. Thermal effects can be limited by pulsing the beam, and most devices have a pulse ratio of 1:5 or 1:4, which means it is on only one-fifth or one-fourth of the time, respectively. A pulse ratio of at least 1:10 is needed to reduce heat production to zero. Nothing significant concerning healing has been reported regarding the chemical effects. Electrical effects may include the attraction of electrophylic metabolites produced during ischemia and pain to be attracted to proteins.

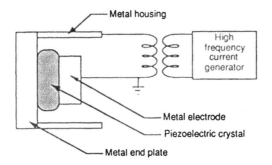

Figure 14.12 A typical ultrasonic transducer and generator. The generator produces an AC current at a frequency equal to the resonant frequency of a quartz crystal. A transducer head placed in front of the crystal vibrates mechanically due to the acoustic vibration of the crystal created when a current is applied across the crystal.

Paul *et al* (1960) used ultrasound to relieve congestion, cleanse necrotic areas, and promote healing. The ultrasonic energy was supplied by a 10-cm round transducer head for 2 to 4 min, three times per week, and up to six times in a treatment course. Water or mineral oil was used to couple the energy to the wound. Dyson and Suckling (1984) discuss the application of low-dosage pulsed ultrasound (0.1 and 0.5 W/cm^2) to accelerate healing. They conclude the main effects of ultrasonic therapy were nonthermal, such as acoustic streaming, which induces changes in diffusion rates, and membrane permeability, which could affect the rate of protein synthesis. Ultrasound is not recommend for infected wounds as it can cause the infection to spread (Wadsworth and Chanmugam 1983).

14.4.7 Future treatments

New methods are constantly being introduced to treat pressure sores. One possible treatment is to inject collagen into pressure sores to reduce stresses and forces and absorb shocks through soft tissue augmentation (Mulder and LaPan 1988). A highly purified bovine dermal collagen has been developed (Collagen Corporation, Palo Alto, CA) and has been used in the feet of diabetics.

Growth factors may also be used to treat pressure sores (Mulder and LaPan 1988). Their effects include fibroblast chemotaxis, RNA synthesis, and increased protein synthesis. Drug delivery systems may also be used to treat pressure sores. Antibiotics, growth factors, or other agents that aid healing may be time-released by dressings or other means.

Functional electrical stimulation (FES) of paralyzed muscled to restore mobility has been used recently by surgeons working with engineers (Buntine and Johnstone 1988). Also known as electrophysiologic orthosis, electroneural prothesis, and electrical bracing, FES has proven to be useful in powering the paralyzed muscles of patients with multiple sclerosis, stroke, and spinal cord injury (Thorsteinsson 1983).

14.4.8 Problems with new treatments

Over the years there have been many claims of new and effective methods of treatment of pressure sores. Fernie and Dornan (1976) pointed out that many new methods for the treatment show increased healing rates initially but are unable to sustain their high performance and are later discarded in favor of older methods. This has caused many "new" cures to be rediscovered and discarded again (Torrance 1983). Allman (1989) states that many treatments are recommended on the basis of uncontrolled studies. Also, many controlled studies fail to ensure the comparability of treatment groups, a serious methodologic flaw.

Fernie and Dornan note that the increased interest generated in staff and patients by the new methods is one variable that has been overlooked by many researchers. They used a placebo device to demonstrate this. The staff using the device was told the device emitted healing EM radiation. The patients that had their treatments supplemented with use of the placebo device showed significant improvements in healing. It was concluded that it is impossible to compare healing rates with control populations due to the added variable of staff interest. To properly compare healing rates for patients using the new treatment with a control population, a placebo must be used on the control population. Cuzzell (1986) has found that a treatment can be successful even if there is no sound physiologic basis for it. The success is attributed to consistent wound assessment and care.

Physicians and therapists should use caution in trying any therapies that have not been verified using a good western scientific model.

14.5 COSTS OF TREATING PRESSURE SORES

The costs directly associated with curing pressure sores have been estimated to exceed $2 billion per year in the United States (Krouskop *et al* 1983). Yearly costs in Great Britain total £70 million ($112 million), and pressure sores lead to a 50% increase in nursing time for each patient with a pressure sore (Barton and Barton 1981). Nawoczenski (1987) reports an estimate of $30,000 as the cost per pressure sore and a five-fold increase in hospitalization costs for those patients that require surgery to close a pressure sore. Wharton *et al* (1988b) analyzed the costs incurred by 50 patients admitted to a spinal cord injury service for the treatment of pressure sores. The average cost including costs of hospital room, surgeon's fees, medications, supplies, and bed rentals was $69,587 during an average stay of 66.5 days. Cervical injuries had a higher average cost ($91,271), followed by thoracic injuries ($56,181) and lumbar injuries ($42,984). Sebern (1986) listed the mean eight-week labor and supply costs for two forms of treatments. The treatment that used a moisture vapor permeable dressing averaged $845, and the treatment that used gauze averaged $1359. The reported average costs per week of treatment vary considerably, and may be a result of including different items such as the cost of the hospital room or surgical fees into the overall cost of treatment.

Numerous lawsuits for compensation of suffering due to development of pressure sores while in the hospital may be on the horizon. Barton and Barton (1981) estimated that the Dept. of Health and Social Security in Great Britain may be liable for greater than £30 million ($48 million). A patient has already been rewarded £2000 ($3200) in one case.

14.6 STUDY QUESTIONS

14.1   Name and define each of the four stages of pressure sore wound healing.
14.2   List and explain factors that affect the rate of healing of pressure sores.
14.3   Name and define types, grades, and colors for pressure sores.
14.4   Explain how you would ensure pressure removal from an ischial sore.
14.5   Explain how you would ensure pressure removal from a trochanteric sore.
14.6   List and define treatments for predisposing factors to improve pressure sore healing.
14.7   Explain the importance of wound debridement.
14.8   Explain how infection of a pressure sore can lead to death.
14.9   Explain the use of flaps in surgical reconstruction of pressure sores.
14.10 Describe advantages and disadvantages of the three major surgical procedures used to close pressure sores.
14.11 Explain the results of electrical stimulation in wound healing.
14.12 Explain the correct statistical design for the testing of a proposed method of improving wound healing.
14.13 Explain how pressure sores can cause death.
14.14 Give and compare the costs of pressure sores in the US and UK.

15

References

Agresti A and Agresti, B 1979 *Statistical Methods for the Social Sciences* (San Francisco: Dellen Publishing)

Agris J and Spira M 1979 Pressure ulcers: Prevention and treatment. *Clin. Symp.* **31** 2–15

Akerblom B 1948 *Standing and Sitting Posture* (Stockholm: Nordiska Bokhandeln)

Akers T K and Gabrielson A L 1984 The effect of high voltage galvanic stimulation on the rate of healing of decubitus ulcers *Biomed. Sci. Instrum.* **20** 99–100

Allman R M, Laprade C A, Noel L B, Walker J M, Moorer C A, Dear M R and Craig R S 1986 Pressure sores among hospitalized patients *Ann. Internal Med.* **105** 337–42

Allman R M 1989 Pressure ulcers among the elderly *N. Engl. J. Med.* **320** 850–3

Alon G, Azaria M and Stein H 1986 Diabetic ulcer healing using high voltage TENS (abstract) *Phys. Ther.* **66** 775

Alvarez O M, Mertz P M, Smerbeck R V and Eaglestein W H 1983 The healing of superficial skin wounds is stimulated by external electrical current *J. Invest. Dermatol.* **81** 144–8

Andersson B J G, Ortengren R, Nachemson A L, Elfstrom G and Breman H 1975 The sitting posture: an electromyographic and discometric study *Orthop. Clin. N. Am.* **6** 105–20

Arnell I 1988 Aggressive and successful prevention of skin breakdown *Todays OR Nurse* **10** 10–4

Babbs C F, Bourland J D, Graber G P, Jones J T and Schoenlein W E 1990 A pressure-sensitive mat for measuring contact pressure distributions of patients lying on hospital beds *Biomed. Instrum. Technol.* **24** 363–70

Bach-y-Rita P, Webster J G, Tompkins W J and Crabb T 1987 Sensory substitution for space gloves and for space robots *Proc. Workshop on Space Telerobotics* 51–7

Bader D L, Gwillim J, Newson T P and Harris J D 1985 Pressure measurement at the patient support interface *Biomechanical measurement in orthopaedic practice* ed M Whittle and J D Harris (Oxford: Oxford University Press) 145–50

Bader D L and Hawken M B 1986 Pressure distribution under the ischium of normal subjects *J. Biomed. Eng.* **8** 353–7

Baer E 1964 Engineering design for plastics *Polymer Science and Engineering Series* (New York: Reinhold)

Bailey B N 1967 *Bedsores* (London: Edward Arnold)

Barbenel J C 1987 The influence of mattress covers and subject weight on interface pressures *Proc. Annu. Conf. Eng. Med. Biol.* **29** 85

Barbenel J C, Ferguson-Pell M W and Kennedy R 1986 Mobility of elderly patients in bed *J. Am. Geriatr. Soc.* **34** 633–6

Bardsley G I, Bell F and Barbenel J C 1976 A technique for monitoring body movements during sleep *Bedsore Biomechanics* ed R M Kenedi, J M Cowden and J T Scales (Baltimore: University Park Press)

Barnes R B 1967 Determination of body temperature by infrared emission *J. Appl. Physiol.* **22** 1143–67

Barron J J, Jacobson W E and Tidd G 1985 Treatment of decubitus ulcers: a new approach *Minn. Med.* **68** 103–6

Barth P W, Le K M, Madsen B L, Ksander G A, Angell J B and Vistnes L M 1984 Pressure profiles in deep tissues *Proc. Annu. Conf. Eng. Med. Biol.* **26** 177

Barton A 1977 Prevention of pressure sores *Nursing Times* **73** 1593–5

Barton A and Barton M 1981 *The Management and Prevention of Pressure Sores* (London: Faber & Faber)

Beebe D J, Denton D D, Webster J G and Radwin R G 1990 A polyimide packaging process for a semiconductor diaphragm tactile sensor *Proc. Annu. Int. Conf. IEEE Eng. Med. Biol. Soc.* **12** 1058–9

Bennett L 1976 Transferring load to flesh: Part VIII. Stasis and stress *Bull. Prosth. Res.* **BPR 10-23** 202–10

Bennett L 1990 Personal communication

Bennett L, Kavner D, Lee B Y and Trainor F S 1979 Shear vs pressure as causative factors in skin blood flow occlusion *Arch. Phys. Med. Rehabil.* **60** 309–14

Bennett L, Kavner D, Lee B Y, Trainor F S and Lewis M L 1984 Skin stress and blood flow in sitting paraplegic patients *Arch. Phys. Med. Rehabil.* **65** 186–90

Bennett L and Lee B Y 1985 Pressure versus shear in pressure sore causation *Chronic Ulcers of the Skin* ed B Y Lee (New York: McGraw-Hill)

Berglund U and Berglund B 1970 Adaptation and recovery in vibrotactile perception *Perceptual and Motor Skills* **30** 843–53

Bergstrom N, Braden B J, Laguzza A and Holman V 1987a The Braden scale for predicting pressure sore risk *Nurs. Res.* **36** 205–10

Bergstrom N, Demuth P J and Braden B J 1987b A clinical trial of the Braden scale for predicting pressure sore risk *Nurs. Clin. N. Am.* **22** 417–28

Betts R P and Duckworth T 1978 A device for measuring plantar pressures under the sole of the foot *Eng. in Med.* **7** 223–8

Black J M and Black S B 1987 Surgical management of pressure ulcers *Nurs. Clin. N. Am.* **22** 429–438

Bobel L M 1987 Nutritional implications in the patient with pressure sores *Nurs. Clin. N. Am.* **22** 379–90

Box G E P, Hunter W G and Hunter J S 1978 *Statistics for Experimenters* (New York: John Wiley & Sons)

Boykin A and Winland-Brown J 1986 Pressure sores nursing management *J. Gerontol. Nurs.* **12** 17–21

Braden B and Bergstrom N 1987 A conceptual schema for the study of the etiology of pressure sores *Rehab. Nurs.* **12** 8–12

Brattgard S-O and Severinsson K 1978 Investigations of pressure, temperature, and humidity in the sitting area in a wheelchair *Biomechanics V1-B. International series on biomechanics* (Baltimore: University Park Press) **2-b** 270–3

Brienza D M, Brubaker C E and Inigo R M 1989a A fiber optic force sensor for automated seating design *Proc. Annu. Conf. Rehabil. Eng. Soc. N. Am.* 232–3

Brienza D M, Chung K C and Inigo R M 1988 Design of a computer aided manufacturing system for custom contained wheelchair cushions *Proc. Int. Conf. Am. Ass. Rehabil. Ther.* 312–3

Brienza D, Gordon J and Thacker J 1989b A comparison of force transducers suitable for an automatic body support contouring system *Proc. Annu. Conf. Rehab. Eng. Soc. N. Am.* 238–9

Brislen W 1982 The Mecabed *Nursing Times* **August 4** 1307–8

Brown A M and Pearcy M J 1986 The effect of water content on the stiffness of seating foams *Prosthet. Orthot. Int.* **10** 149–52

Brown M and Gogia P 1986 Effects of high voltage galvanic stimulation on wound healing (abstract) *Phys. Ther.* **66** 748

Bryan C S, Dew C E and Reynolds K L 1983 Bacteremia associated with decubitus ulcers *Arch. Intern. Med.* **143** 2093–5

Bunch W H and Keagy R D 1976 *Principles of Orthotic Treatment* (St. Louis: Mosby)

Buntine J A and Johnstone B R 1988 The contributions of plastic surgery to care of the spinal cord injured patient *Paraplegia* 26 87–93

Burn T G, Ferguson–Pell M W and Hurwitz D 1985 Pressure relief monitoring device *Proc. Annu. Conf. Rehab. Eng. Soc. N. Am.* 42–4

Burnett D S 1987 Finite element analysis from concepts to applications (Reading, MA: Addison-Wesley)

Burr A H 1981 *Mechanical analysis and design* (Elsevier: New York) 486–97

Carley P J and Wainapel S 1985 A clinical application of electrotherapy for accelerating wound healing *Proc. Annu. Conf. Rehab. Eng. Soc. N. Am.* 8 114–6

Chawla J C, Andrews B and Bar C 1978–79 Using warning devices to improve pressure-relief training *Paraplegia* 16 413–9

Cherry G W, Ryan T J and Ellis J P 1974 Decreased fibrinolysis in reperfused ischaemic tissue *Thrombo. Diasthes. Haemorrh.* 32 659–71

Chizek H J, Selwan P M and Merat F L 1985 A foot pressure sensor for use in lower extremity neuroprosthetic development *Proc. Annu. Conf. Rehabil. Eng. Soc. N. Am.* 8 379–80

Chow W W and Odell E I 1978 Deformations and stresses in soft body tissues of a sitting person *Trans. ASME, J. Biomech. Eng.* 100 79–87

Christensen D A 1980 A review of current optical techniques for biomedical physical measurements *Physical Sensors for Biomedical Applications* ed M R Neuman, D G Fleming and W H Ko (Boca Raton, FL: CRC Press)

Chung K C 1987 *Tissue Contour and Interface Pressure on Wheelchair Cushions* Ph D Dissertation, University of Virginia

Chung K C, DiNello A M and McLaurin C A 1987 Comparative evaluation of pressure distribution on foams and contoured cushions *Proc. Annu. Conf. Rehabil. Eng. Soc. N. Am.* 10 323–5

Clark M O, Barbenel J C, Jordan M M and Nicol S M 1978 Pressure sores *Nursing Times* 74 363–6

Clark R A F 1988 Overview and general considerations of wound repair *The Molecular and Cellular Biology of Wound Repair* ed R A F Clark and P M Henson (New York: Plenum)

Claus-Walker J and Halstead L S 1982a Metabolic and endocrine changes in spinal cord injury: III. Less quanta of sensory input plus bedrest and illness *Arch. Phys. Med. Rehabil.* 63 628–31

Claus-Walker J and Halstead L S 1982b Metabolic and endocrine changes in spinal cord injury: IV. Compounded neurologic dysfunctions *Arch. Phys. Med. Rehabil.* 63 632–8

Cochran G V B and Palmieri V 1981 Development of a modular wheelchair cushion *Paraplegia News* **35** 45

Conine T A, Choi A K M and Lim R 1989 The user-friendliness of protective support surfaces in prevention of pressure sores *Rehabil. Nurs.* **14** 261–3

Conry T F and Seireg A 1971 A mathematical programming method for design of elastic bodies in contact *J. Appl. Mech.* **93** 387–92

Constantian M B 1980 Historical note *Pressure Ulcers: Principles and Techniques of Management* ed M B Constantian (Boston: Little, Brown) 3–6

Constantian M B and Jackson H S 1980d The ischial ulcer *Pressure Ulcers: Principles and Techniques of Management* ed M B Constantian (Boston: Little, Brown) 215–46

Constantian M B and Jackson H S 1980a Biology and care of the pressure ulcer wound *Pressure Ulcers: Principles and Techniques of Management* ed M B Constantian (Boston: Little, Brown) 69–100

Constantian M B and Jackson H S 1980b The complex problem *Pressure Ulcers: Principles and Techniques of Management* ed M B Constantian (Boston: Little, Brown) 261–89

Constantian M B and Jackson H S 1980c The sacral ulcer *Pressure Ulcers: Principles and Techniques of Management* ed M B Constantian (Boston: Little, Brown) 149–84

Cooper D G and Hawkes E 1984 The meru shapeable matrix support surface for children and adults *Proc 2nd Intl Conf Rehabil Eng* 475–6

Cooper D, Roxborough L, Fife S, Tredwell S and Neen D 1986 Pressure monitoring chair *Proc. Annu. Conf. Res. Eng. Soc. N. Am.* **9** 390–3

Copeland-Fields L D and Hoshiko B R 1989 Clinical validation of Braden and Bergstrom's conceptual schema of pressure sore risk factors *Rehabil. Nurs.* **14** 257–60

Could B S 1968 Hydroproline and the Metabolism of Collagen *Treatise on Collagen* (New York: Academic Press)

Crenshaw R P and Vistnes L M 1989 A decade of pressure sore research: 1977–1987 *J. Rehab. and Dev.* **26** 63–74

Crewe R 1983 The role of the occupational therapist in pressure sore prevention In: *Pressure sores* (ed. J C Barbanel, C D Forbes and G D Lowe) (Macmillan: London) 121–32

Cumming W T, Tompkins W J, Jones R M and Margolis S A 1986 Microprocessor-based weight shift monitors for paraplegic patients *Arch. Phys. Med. Rehabil.* **67** 172–4

Cuzzell J Z 1986 Wound care forum: Reader's remedies for pressure sores *Am. J. Nurs.* **86** 923–4

Daniel R K, Priest D L and Wheatley D C 1981 Etiologic factors in pressure sores: An experimental model *Arch. Phys. Med. Rehabil.* **62** 492–8

Dario P and De Rossi D 1985 Tactile sensors and the gripping challenge *IEEE Spectrum* 22 (8) 46–52

David J A, Chapman R G, Chapman E J and Lockett B *An investigation of the current methods used in nursing for the care of patients with established pressure sores* Guildford: Nursing Practice Research Research Unit, University of Surrey, 1983

Denne W A 1981 The "hammock" effect in wheelchair cushion covers *Paraplegia* 19 38–42

Department of Veterans Affairs 1990 Choosing a wheelchair system *J. Rehabil. Res. Dev., Clin. Suppl. #2*, 26 1–118

Derrington C E, Newton W B, Gragg JE and Brice-Heames K V 1982 *An Integrated Silicon Pressure Sensor Using a Four Terminal Shear Stress Sensitive Piezoresistive Strain Gauge* (Phoenix, AZ: Motorola Inc.)

Diffrient N, Tilley A R and Bardagjy J C 1974 *Humanscale 1/2/3* (Cambridge, MA: MIT)

Dinsdale S M 1973 Decubitus ulcers in swine: light and electron microscopy study of pathogenesis *Arch. Phys. Med. Rehabil.* 54 51–6

Dinsdale S M 1974 Decubitus ulcers, role of pressure and friction in pressure sores *Arch. Phys. Med. Rehabil.* 55 147–52

Doebelin E O 1990 *Measurement Systems, Application and Design* 4th ed. (New York: McGraw-Hill)

Dyson M and Suckling J 1984 Stimulation of tissue repair by ultrasound: a survey of the mechanisms involved *Physiotherapy* 64 105–8

Ek A C and Bowman G 1982 A descriptive study of pressure sores: The prevalence of pressure sores and the characteristics of patients *J. Advanced Nurs.* 7 51–7

El-Toraei I and Chung B 1977 The management of pressure sores *J. Dermatol. Surg. Oncol.* 3 507–11

Eltorai I 1981 Hyperbaric oxygen in the management of pressure sores in patients with injuries to the spinal cord *J. Dermatol. Surg. Oncol.* 7 737–40

Esashi M, Komatsu H, Matsuo T, Takahashi M, Takishima T, Imabayashi K and Ozawa H 1982 Fabrication of catheter-tip and sidewall miniature pressure sensors *IEEE Trans. Elec. Dev.* ED-29 57–9

Exton-Smith A N and Sherwin R W 1961 The prevention of pressure sores: significance of spontaneous bodily movements *Lancet* 1 124–6

Fan L S, White R M and Muller R S 1984 A mutual capacitive normal- and shear-sensitive tactile sensor *International Electron Devices Meeting* 220–2

Feeder J A and Kloth L C 1985 Acceleration of healing with high voltage pulsating direct current (abstract) *Physical Ther.* 65 741

Ferguson-Pell M W 1980 Design criteria for the measurement of pressure at body/support interfaces *Eng Med* **9** 209–14

Ferguson-Pell M W 1990 Seat cushion selection *J. Rehab. Res. Dev.* **Clin. Suppl. No. 2** 49–73

Ferguson-Pell M W, Bell F and Evans J H 1976 Interface pressure sensors: existing devices, their suitability and limitations *Bedsore Biomechanics* ed R M Kenedi, J M Cowden and J T Scales (Baltimore: University Park Press)

Ferguson-Pell M, Cochran G V B, Palmieri V R and Brunski J B 1986 Development of a modular wheelchair cushion for spinal cord injured persons *J. Rehabil. Res. Dev.* **23** 63–76

Ferguson-Pell M and Hagisawa S 1987 Biochemical changes in sweat following prolonged ischemia *Proc. Annu. Conf. Eng. Med. Biol.* **29** 90

Ferguson-Pell M W, Wilkie I C, Reswick J B and Barbenel J C 1980 Pressure sore prevention for the wheelchair-bound spinal injury patient *Paraplegia* **18** 42–51

Fernandez S 1987 Physiotherapy prevention and treatment of pressure sores *Physiotherapy* **73** 450–4

Fernie G R and Dornan J 1976 The problems of clinical trials with new systems for preventing or healing decubiti *Bedsore Biomechanics* ed R M Kenedi, J M Cowden and J T Scales (Baltimore: University Park Press)

Fisher S V and Kosiak M 1979 Pressure distribution and skin temperature effect of the ROHO wheelchair balloon cushion *Arch. Phys. Med..Rehabil.* **60** 70–1

Fisher S V and Patterson R P 1983 Long term pressure recording under the.ischial tuberosities of tetraplegics *Paraplegia* **21** 99–196

Fisher S V, Szymke T E, Apte S Y and Kosiak M 1978 Wheelchair cushion effect on skin temperature *Arch. Phys. Med. Rehabil.* **59** 68–72

Flam E 1987 Optimum skin aeration in pressure sore management *Proc. Annu. Conf. Eng. Med. Biol.* **29** 84

Floyd W F and Roberts D F 1958 Anatomical and physiological principles in chair and table design *Ergonomics* **2** 1–16

Fowler E M 1987 Equipment and products used in management and treatment of pressure ulcers *Nurs. Clin. N. Am.* **22** 449–61

Francis J E, Roggli R, Love T J and Robinson C P 1979 Thermography as a means of blood perfusion measurement *Trans. ASME, J. Biomech. Eng.* **101** 246–9

Francis Jones H W and Evans A 1978 The polystyrene vacuum wheelchair cushion *Paraplegia* **16** 420–2

Frisbie J H 1986 Wound healing in acute spinal cord injury: effect of anticoagulation *Arch. Phys. Med. Rehabil.* **67** 311–3

Ganong W F 1989 *Review of Medical Physiology* (San Mateo: Appleton & Lange)

Gardner W J and Anderson R M 1954 The alternating pressure pad: an aid to the proper handling of decubitus ulcers *Arch. Phys. Med. Rehabil.* **35** 578–80

Garfin S R, Pye S A, Hargens A R and Akeson W H 1980 Surface pressure distribution of the Human body in the recumbent position *Arch. Phys. Med. Rehabil.* **61** 409–13

Gerson L W 1975 Incidence of pressure sores in active treatment hospitals *Int. J. Nurs. Studies* **2** 201–4

Giallorenzi T G, Bucaro J A, Dandrige A, Sigel G H, Cole J H, Rashleigh SC and Priest R G 1982 Optical fiber sensor technology *IEEE J. Quantum Electron.* **QE-18** (4) 626–65

Gibson D A and Wilkins K E 1975 The management of spinal deformities in duchenne muscular dystrophy *Clin. Orthopaedics Related Res.* **108** 41–51

Gilsdorf P, Patterson R, Fisher S and Appel N 1990 Sitting forces and wheelchair mechanics *J. Rehabil. Res. Dev.* **27** 239–46

Giltvedt J, Sira A and Helme P 1984 Pulse multifrequency photoplethysmograph *Med. Biol. Eng. Comput.* **22** 212–15

Ginpil F, Milner M and Rang M 1984 Towards an economical postural seating system using an adjustable mesh *Proc. 2nd Int. Conf. Rehabil. Eng.* 489–90

Goldstone L A and Goldstone J 1982 The Norton score: an early warning of pressure sores? *J. Adv. Nurs.* **7** 419–26

Goller H, Lewis D W and McLaughlin R E 1971 Thermographic studies of human skin subjected to localized pressure *Am. J. Roentgenol. Radium Ther. Nucl. Med.* **113** 749–54

Gosnell D J 1973 An assessment tool to identify pressure sores *Nurs. Res.* **22** 55–7

Gosnell D J 1987 Assessment and evaluation of pressure sores *Nurs. Clin. N. Am.* **22** 399–416

Grandjean E, Boni A and Kretzschmar H 1969 The development of a rest chair profile for healthy and notalgic people *Sitting posture* ed E. Grandjean (London: Taylor and Francis) 193–201

Green M F 1976 Skin care: A GP's guide to safe effective treatment of bedsores *Mod. Geriat.* **6** 38–42

Griffith B H 1963 Advances in treatment of decubitus ulcers *Surg. Clin. N. Am.* **43** 245–60

Grimby G and Widerstad-Lossing I 1989 Comparison of high- and low-frequency muscle stimulators *Arch. Phys. Med. Rehabil.* **70** 835–8

Grip J C and Merbitz C T 1986 Wheelchair-based mobile measurement of behavior for pressure sore prevention *Comput. Methods Programs Biomed.* **22** 137–44

Gross D R, Beasley C W and Copcutt B G 1982 Real time determination of 81Rb/81mKr ratio to measure blood flow in subcutaneous tissues under pressure *Proc. Annu. Conf. Rehabil. Eng. Soc. N. Am.* **5** 83

Groth E K 1942 Klinische Beobachtungen und experimentelle Studien uber die Enstehung des Dekubitus *Acta. Chir. Scand.* **87** 198–200

Guckel H and Burns D W 1984 Planar processed polysilicon sealed cavities for pressure transducer arrays *Proc. IEEE IEDM* 223–5

Guttmann L 1973 *Spinal cord injuries: comprehensive management and research* (Oxford: Blackwell Scientific)

Guttmann L 1976 The prevention and treatment of pressure sores *Bedsore Biomechanics* ed R M Kenedi, J M Cowden and J T Scales (Baltimore: University Park Press)

Guyton A C 1981 *Basic human physiology* 3rd ed. (Philadelphia: W B Saunders)

Hagisawa S, Ferguson-Pell M W, Palmieri V R and Cochran G V B 1988 Pressure sores: a biochemical test for early detection of tissue damage *Arch. Phys. Med. Rehabil.* **69** 668–71

Hagner D 1988 Metal strain gages *Tactile Sensors for Robotics and Medicine* ed J G Webster (New York: John Wiley & Sons)

Hahm G and Webster J G 1989 *A miniature optoelectronic force sensor* University of Wisconsin, Dept. Elec. Comp. Eng.

Hall O C and Brand P W 1979 The etiology of the neuropathic plantar ulcer *J. Am. Podiatry Assoc.* **69** 173–7

Hargens A R 1986 *Tissue Nutrition and Viability* ed A R Hargens (Springer-Verlag: New York)

Hargest T S 1976 Problems of patient support:the air fluidised bed as a solution *Bedsore Biomechanics* ed R M Kenedi, J M Cowden and J T Scales (Baltimore: University Park Press)

Harman J W 1948 Significance of local vascular phenomena in production of ischemic necrosis in skeletal muscle *Am. J. Pathol.* **24** 625–41

Harper H, Rodwell V and Mayes P 1979 *Review of Physiological Chemistry* (Los Altos, CA: Lange Medical Publications)

Harries R 1987 Personal experience of pressure sores *Physiotherapy* **73** 448–50

Harris R I 1947 *Army foot survey. An investigation of foot ailments in Canadian soldiers, No. 1574* (Ottawa, Canada: National Research Council)

Hassard G H, Conry J and Rice W 1971 Bucket seat for control of decubitus ulcers *Arch. Phys. Med. Rehabil.* **52** 481–4

Hayne C R 1984 Pulsed high frequency energy—its place in physiotherapy *Physiotherapy* **70** 459–66

Heistad D D and Abboud F M 1974 Factors that influence blood flow in skeletal muscle and skin *Anesthesiology* **41** 139–56

Hertzman A B 1938 The blood supply of various skin areas as estimated by the photoelectric plethysmograph *Am. J. Physiol.* **124** 328–40

Hobson D A 1989 Comparative effects of posture on pressure distribution at the body-seat interface *Proc. Annu. Conf. Rehabil. Eng. Soc. N. Am.* **12** 83–4

Hobson D A and Tooms R 1981 The foam-in-place seating system results of toxicity studies *Proc. Annu. Conf. Rehabil. Eng. Soc. N. Am.* **4** 45–8

Hobson D A, Taylor S J and Shaw C G 1986 Bead matrix insert system- a follow-up clinical report *Proc. Annu. Conf. Rehabil. Eng. Soc. N. Am.* **9** 381–3

Hole J W 1987 *Human Anatomy and Physiology* (Dubuque: Wm. C. Brown)

Holley L K, Long J, Stewart J and Jones R F 1979 A new pressure measuring system for cushions and beds—with a review of the literature *Paraplegia* **17** 461–74

Holloway G A, Daly C H, Kennedy D and Chimoskey J 1976 Effects of external pressure loading on human skin blood flow measured by ^{133}Xe clearance *J. Appl. Physiol.* **40** 597–600

Houle R J 1969 Evaluation of seat devices designed to prevent ischemic ulcers in paraplegic patients *Arch. Phys. Med. Rehabil.* **50** 587–94

Hughes J, Kriss S and Kleneman L 1987 A clinician's view of foot pressure: a comparison of 3 different methods of measurements *Foot & Ankle* **7** 277–84

Hunt R K 1978 Anaerobic metabolism and wound healing: An hypothesis for the initiation and cessation of collagen synthesis in wounds *Am. J. Surg.* **135** 328

Hunt T 1979 Disorders of repair and their treatment *Fundamentals of Wound Management* ed T K Hunt and J E Dunphy (New York: Appleton-Century-Crofts)

Hunt T K and Van Winkle W 1979 Normal repair *Fundamentals of Wound Management* ed T K Hunt and J E Dunphy (New York: Appleton-Century-Crofts)

Hunter J A, McVittie E and Comaish J S 1974 Light and electron microscopic studies of physical injury to the skin. II. Friction *Br. J. Dermatol.* **90** 491–9

Hunter K M, Ko V and Suggitt R 1983 Bed turner *Proc. Annu. Conf. Rehabil. Eng. Soc. N. Am.* **6** 286–8

Husain T 1953 An experimental study of some pressure effects on tissues, with reference to the bed-sore problem *J. Path. Bact.* **LXVI** 347–58

Hynecek J. 1975 Miniature pressure transducer for biomedical applications *Indwelling and Implantable Pressure Transducers* ed D G Fleming, W H Ko and M R Neuman (Cleveland, OH: CRC Press)

Interlink 1987 *Application note IL-02: The Interlink force sensing resistor kit* Santa Barbara, CA

Irvine R E, Memon A H and Shera A S 1961 Norethandrolone and prevention of pressure sores *Lancet* December 16 1333

Isherwood P A 1976 The use of chemically dried air in the treatment of chronic ulcers *Bedsore Biomechanics* ed R M Kenedi, J M Cowden and J T Scales (Baltimore: University Park Press)

Jaeger R J and Marlantis S 1983 Pressure relief training device *Proc. Annu. Conf. Eng. Med. Biol.* 25 148

Jamsa R, Wilander L, Siwe H and Odman S 1990 Fibre-optic force recording of mechanical contact between teeth *Med. Biol. Eng. Comp.* 28 89–91

Jaros L A 1987 *The SPIRAL pressure monitor* Dept Physical Medicine Rehabilitation, Univ. Michigan, Ann Arbor, MI

Jaros L, Kett R and Levine S 1986 A computer interface for a seating pressure evaluator *Proc. Annu. Conf. Rehabil. Eng. Soc. N. Am.* 9 419–21

Jaros L A, Levine S P, Kett R L and Koester D J 1988 The spiral pressure monitor *ICAART 88—Montreal* 308–9

Jeneid P 1976 Static and dynamic support systems—pressure differences on the body *Bedsore Biomechanics* ed R M Kenedi, J M Cowden and J T Scales (Baltimore: University Park Press)

Jenkins D H R, Forster I W, McKibbin B and Ralis Z 1977 Induction of tendon and ligament formation by carbon implants *J. Bone Joint Surg.* 59B 53–7

Jensen T 1990 *Calibrating and using Interlink sensors* Dept. Elec. Comp. Eng., Univ. Wisconsin, Madison WI

Jercinovic A, Bobanovic F, Rebersek S, Karba R, Stefanovska A and Vodovnik L 1989 Endogenous potentials of injured skin *Electromagnetic Fields and Biomembranes* (Pleven: Second International School)

Jordan M M and Clark M O 1977 *Report on the Incidence of Pressure Sores in the Patient Community of the Greater Glasgow Health Board Area on 21 January 1976* (Glasgow: University of Strathclyde Bioengineering Unit and The Greater Glasgow Health Health Board)

Jordan M M and Nicol S M 1977 *Report on the Incidence of Pressure Sores in the Patient Community of the Borders Health Board Area on 13th October 1976* (Glasgow: University of Strathclyde Bioengineering Unit and Borders Health Board)

Joyner A W 1988 Weight transfer wheelchair seat *ICAART 88-MONTREAL* 254–5

Kaada B 1983 Promoted healing of chronic ulceration by transcutaneous nerve stimulation (TNS) *VASA* **12** 262–3

Kaczmarek K A, Webster J G, Bach-y-Rita P and Tompkins W J 1991 Electrotactile and vibrotactile displays for sensory substitution *IEEE Trans. Biomed. Eng.* **38** January

Kadaba M P, Ferguson-Pell M W, Palmieri V R and Cochran G V B 1984 Ultrasound mapping of the buttock cushion interface contour *Arch. Phys. Med. Rehabil.* **65** 467–9

Kahn J 1987 *Principles and Practice of Electrotherapy* (New York: Churchill Livingstone)

Kambic H E, Reyes E, Manning T and Reger S I 1990 Orientation and biomechanical properties of dc electrically stimulated and healed pigskin pressure sores *Advances in External Control of Human Extremities X* ed D Popovic (Belgrade: Yugoslav Committee for Electronics and Automation)

Katzung B 1989 *Basic and Clinical Pharmacology* (Los Altos, CA: Appleton and Lange)

Kauer C and Sonsino G 1986 The need for skin and muscle saving techniques in the repair of decubitus ulcers *Scand. J. Plast. Reconstr. Surg.* **20** 129–31

Keane F X 1978-79 The minimum physiological mobility requirement for man supported on a soft surface *Paraplegia* **16** 383–9

Kenedi R M, Cowden J M and Scales J T (eds) 1976 *Bedsore Biomechanics* (Baltimore: University Park Press)

Kennedy E J 1986 *Spinal Cord Injury: The Facts and Figures* (Birmingham, AL: University of Alabama)

Kett R L and Levine S P 1987 A dynamic model of tissue deflection in a seated individual *Proc. Annu. Conf. Rehab. Eng. Soc. N. Am.* **10** 524–6

Kett R L, Levine S P, Bowers L D, Brooks S V and Cederna P S 1986 Dynamic effects of functional electrical stimulation on seating interface pressure and tissue shape. *Proc. Annu. Conf. Rehabil. Eng. Soc. N. Am.* **9** 313–5

Kett R L, Levine S P, Wilson B A and Gross M D 1988 Ischial blood flow in the skin of seated SCI individuals during electrical muscle stimulation *ICAART88 - Montreal* 324–5

Key A G, Manley M T and Wakefield E 1978 Pressure redistribution in wheelchair cushion for paraplegics: its application and evaluation *Paraplegia* **16** 403–12

Kirschner F D 1986 *Tactile stimulation as a substitute for vision for the blind* PhD dissertation, Ohio State University

Klein R M and Fowler R S 1981 Pressure relief training device: the microcalculator *Arch. Phys. Med. Rehabil.* **62** 500–1

Kleitman N 1939 *Sleep and wakefulness* (Chicago: University of Chicago Press)

Ko W, Bao M-H and Hong Y-D 1982 A high-sensitivity integrated-circuit capacitive pressure transducer *IEEE Trans. Elec. Dev.* **ED-29** 48–51

Ko W, Hynecek J, Boettcher S F 1979 Development of a miniature pressure transducer for biomedical applications *IEEE Trans. Elec. Dev.* **ED-26** 1896–905

Koreska J , Gibson D A and Albisser A M 1976 Structural support system for unstable spines Biomechanics V-A. *International series on biomechanics* (Baltimore: University Park Press) **1-A** 474–83

Kosiak M 1959 Etiology and pathology of ischemic ulcers *Arch. Phys. Med. Rehabil.* **39** 62–9

Kosiak M 1961 Etiology decubitus ulcers *Arch. Phys. Med. Rehabil.* **41** 19–29

Kosiak M 1976 A mechanical resting surface: its effect on pressure distribution *Arch. Phys. Med. Rehabil.* **57** 481–4

Kosiak M, Kubicek W G, Olson M,Danz J N and Kottke F J 1958 Evaluation of pressure as a factor in the production of ischial ulcers *Arch. Phys. Med. Rehabil.* **39** 623–8

Kostunik J P and Fernie G 1985 Pressure sores in elderly patients (editorial) *J. Bone Joint Surg.* **67-B** 1–2

Kovach M W 1985 *Design consideration for construction of a vibrotactile array* Master of Science Thesis, Ohio State University

Krebs M A 1984 A comparison of the pressure relief characteristics of two air flotation beds *Proc. 2nd Int. Conf. Rehabil. Eng. OTTAWA* 461–2

Krouskop T A 1983 A synthesis of the factors that contribute to pressure sore formation *Med. Hypoth.* **11** 255–67

Krouskop T, Krebs M, Herszkowicz I and Garber S 1985b Effectiveness of mattress overlays in reducing interface pressure during recumbency *J. Rehabil. Res. Dev.* **22** 7–10

Krouskop T A, Noble P, Brown J and Marburger R 1986 Factors affecting the pressure-distributing properties of form mattress overlays *J. Rehabil. Res. Dev.* **23** 33–9

Krouskop T, Noble P, Garber S and Brown J 1985a Posture, comfort and interface pressures while recumbent *Proc. Annu. Conf. Eng. Med. Biol.* **27** 140

Krouskop T A, Noble P C, Garber S L and Spencer W A 1983 The effectiveness of preventive management in reducing the occurrence of pressure sores *J. Rehab. Res. Dev.* **20** 74–83

Krouskop T A, Williams R, Kerbs M and Garber S 1985 Effectiveness of mattress overlays in reducing interface pressures during recumbency *J. Rehabil. Res. Dev.* **22** 7–10

Krouskop T A, Williams R, Noble P and Brown J 1986 Inflation pressure effect on performance of air-filled wheelchair cushions *Arch. Phys. Med. Rehabil.* **67** 126–8

Laing E and Walton S K 1979 Report on the clinical evaluation of the Simpson-Edinburgh low pressure air bed (Edinburgh: Bio-engineering unit, Princess Margaret Rose Hospital)

Laliberte B J, Masiello M P, Kadaba M P and Cochran G V B 1985 Mapping the deformation and pressure contours at the buttock–cushion interface *Proc. Annu. Conf. Rehabil. Eng. Soc. N. Am.* **8** 344–6

Landis E M 1930 Micro-injection studies of capillary blood pressure in human skin *Heart* **15** 209–28

Larson J and Risberg B 1977 Ischemia-induced changes in tissue fibrinolysis in human legs *Bibl. Anat.* **15** 556–8

Le K M, Madsen B L, Barth P W, Ksander G A, Angell J B and Vistnes L M 1984 An in-depth look at pressure sores using monolithic silicon pressure sensors *Plastic Reconstr. Surg.* **74** 745–54

Lee B Y 1985 *Chronic Ulcers of the Skin* (New York: McGraw-Hill)

Lee B Y, Trainor F S, Kavner D, Crisologo J A, Shaw W and Madden J L 1979 Assessment of the healing potentials of ulcers of the skin by photoplethysmography *Surg. Gynecol. Obstet.* **148** 233–9

Lee, Y S and Wise K D 1982 A batch -fabricated silicon capacitive pressure transducer with low temperature sensitivity *IEEE Trans. Elec. Dev.* **ED-29** 42–8

Levine J M, Simpson M and McDonald R 1989a Pressure sores: a plan for primary care prevention *Geriatrics* **44** 75–90

Levine S P, Kett R L, Cederna P S and Brooks S V 1990a Electric muscle stimulation for pressure sore prevention: tissue shape variation *Arch. Phys. Med. Rehabil.* **71** 210–5

Levine S P, Kett R L, Cederna P S, Bowers L D and Brooks S V 1989b Electrical muscle stimulation for pressure variation at the seating interface *J. Rehabil. Res. Dev.* **26** 1–8

Levine S P, Kett R L, Finestone H M, Carlson G A and Chizinsky K A 1989c Electrical muscle stimulation for the prevention of pressure sores: clinical trials *Rehabil. R & D Progress Reports 1989, J. Rehabil. Res. Dev.* **26** (Suppl.) 288

Levine S P, Kett R L, Gross M D, Wilson B A, Cederna P S and Juni J E 1990b Blood flow in the gluteus maximus of seated individuals during electrical muscle stimulation *Arch. Phys. Med. Rehabil.* **71** 682–6

Lewis D W and Nourse W B 1978 A device designed to approximate shear forces on human skin *Bull. Prosthetics Res.* **10-30** 36–46

Lewis R C 1977 *Handbook of Traction, Casting and Splinting Techniques* (Philadelphia: J B Lippincott)

Lim R, Sirett R, Conine T A and Daechsel D 1988 Clinical trial of foam cushions in the prevention of decubitus ulcers in elderly patients *J. Rehab. Res. Dev.* **25** 19–26

Lindan O 1961 Etiology of decubitus ulcers: An experimental study *Arch. Phys. Med. Rehabil.* **42** 774–83

Lomas C 1988 Wound care. From theory to practice *Nursing Times* **84** 63–6

Love T J and Lindsted R D 1975 Theoretical basis for use of skin temperature as a plethysmographic indicator ASME paper no. 75–WA/Bio–6

Lowthian P 1979 Pressure sore prevalence *Nursing Times* **75** 358–60

Maalej N, Bhat S, Zhu H, Webster J G, Tompkins W J, Wertsch J J and Bach-y-Rita P 1988 A conductive polymer pressure sensor *Proc. Annu. Int. Conf. IEEE Eng. Med. Biol. Soc.* **10** 770–1

Maalej N and Webster J G 1988 A miniature electrooptical force transducer *IEEE Trans. Biomed. Eng.* **35** 93–8

Maalej N, Zhu H, Webster J G, Tompkins W J, Wertsch J J and Bach-y-Rita P 1987 Pressure monitoring under insensate feet *Proc. Annu. Int. Conf. IEEE Eng. Med. Biol. Soc.* **9** 1823–4

Mahanty S D and Roemer R B 1980 Thermal and circulatory response of tissue to localized pressure application: a mathematical model *Arch. Phys. Med. Rehabil.* **61** 335–40

Mahanty S D, Roemer R B and Meisel H 1981 Thermal response of paraplegic skin to the application of localized pressure *Arch. Phys. Med. Rehabil.* **62** 608–11

Maklebust J 1987 Pressure ulcers: Etiology and prevention *Nurs. Clin. N. Am.* **22** 359–75

Malament I B, Dunn M E and Davis R 1975 Pressure sores: an operant conditioning approach to prevention *Arch. Phys. Med. Rehabil.* **56** 161–5

Manley M T 1978 Incidence, contributory factors and costs of pressure sores *S. Afr. Med. J.* **53** 217–22

Marshall R S 1971 Cold therapy in the treatment of pressure sores *Physiotherapy* **57** 372–3

Mason J L and Mackay N A M 1976 Pain sensations associated with electrocutaneous stimulation *IEEE Trans. Biomed. Eng.* **BME-23** 405–9

Mawson A R, Neville P and Winchester Y 1988 Risk factors for early occurring pressure ulcers following spinal cord injury *Am. J. Phys. Med. Rehab.* **67** 123–7

Mayer L 1936 Further studies of fixed paralytic pelvic obliquity *J. Bone Joint Surg.* **18** 87–100

Mayo-Smith W 1980 Pressure measurements: device for obtaining readouts from multiple sites using pneumatic transducers *Arch. Phys. Med. Rehabil.* **61** 460–1

McDougall A 1976 Clinical aspects of bed sore prevention and treatment *Bedsore Biomechanics* ed R M Kenedi, J M Cowden and J T Scales (Baltimore: University Park Press)

McGovern T, Magnano S, Filip K, Stewart T and Reger S 1987 The comparison of support interface pressure measuring systems *Proc. Annu. Conf. Rehabil. Eng. Soc. N. Am.* **10** 533–5

McGovern T F and Reger S I 1987 Digital interface pressure evaluator *Proc. Annu. Conf. Rehabil. Eng. Soc. N. Am.* **10** 305–7

McGovern T F, Reger S I, Snyder E N and Sauer B L 1988 A new technique for custom contoured body supports *Proc. Int. Conf. Am. Ass. Reh. Ther.* 306–7

McGraw-Hill Encyclopedia of Science and Technology, 6th Edition 1987 *Photoelasticity* (New York: McGraw-Hill)

McKenzie M W and Rogers J E 1973 Use of trunk supports for severely paralyzed people *Am. J. Occup. Ther.* **27** 147–8

Mechanic H F and Perkins B A 1988 Preventing tissue trauma *Dimens. Crit. Care Nurs.* **7** 210–8

Meinecke E A and Clark R C 1973 Mechanical properties of polymeric foams (Westport, CT: Technomic Publishing)

Merbitz C T, King R B, Bleiberg J and Grip J 1985a Wheelchair push-ups: measuring pressure relief frequency *Arch. Phys. Med. Rehabil.* **66** 10–34

Merbitz C T, Rosemarie B K, Bleiberg J and Grip J C 1985b Wheelchair push-ups: measuring pressure relief frequency *Arch. Phys. Med. Rehabil.* **66** 433–8

Miller G E and Seale J 1981 Lymphatic clearance during compressive loading *Lymphology* **14** 161–6

Miller M E and Sachs M L 1974 *About bedsores: What you need to know to help prevent and treat them* (Philadelphia: Lippincott)

Minns R J, Sutton R A, Duffus A and Mattinson R 1984 Underseat pressure distribution in the sitting spinal injury patient. *Paraplegia* **22** 297–304

Moe J H, Winter R B, Bradford D S and Lomstein J E 1978 *Scoliosis and Other Spinal Deformities* (Philadelphia: Saunders)

Montgomery D C 1976 *Design and Analysis of Experiments* (New York: John Wiley & Sons)

Mooney V, Einbund J J, Rogers J E and Stauffer E S 1971 Comparison of pressure qualities in seat cushions *Bull. Pros. Res.* **10-15** 129–43

Mubarak S J, Hargens A R, Owen C A, Garetto L P and Akeson W H 1976 The wick catheter technique for measurement of intramuscular pressure *J. Bone Joint Surg.* **58-A** (7) 1016–20

Mulder G D and LaPan M 1988 Decubitus ulcers: update on new approaches to treatment *Geriatrics* **43** 37–39, 44–45, 49–50

Murray Y 1988 Wound care. Tradition rather than cure? *Nursing Times* **84** 75–80

Muzet A, Becht J, Jacquot P and Koenig P 1972 A technique for recording human body posture during sleep *Psychophysiology* **9** 660–2

Natow A B 1983 Nutrition in the prevention and treatment of decubitus ulcers *Aspen Systems Corp.* 39–44

Nawoczenski D A 1987 Pressure sores: prevention and management *Spinal Cord Injury: Concepts and Management Approaches* ed L E Buchanan and D A Nawoczenski (Baltimore: Williams & Wilkins)

Nemethy G and Scheraga H A 1986 Stabilization of collagen fibrils by hydroxyproline *Biochemistry* **25** 3184–8

Neth D C, McGovern T F and Reger S I 1989 Computer aided measurement and fabrication of contoured body supports *Proc. Annu. Conf. Rehabil. Eng. Soc. N. Am.* **12** 234–5

Neuman M R, Berec A and O'Connor E 1984 Capacitive sensors for measuring finger and thumb tip forces *Proc. Annu. Int. Conf. IEEE Eng. Med. Biol. Soc.* **6** 436–9

Newman P and Davis N H 1981 Thermography as a predictor of sacral pressure sores *Age and Ageing* **10** 14–8

Newson T P, Pearcy M J and Rolfe P 1981 Skin surface PO_2 measurements and effect of externally applied pressure *Arch. Phys. Med. Rehabil.* **62** 390–2

Newson T P and Rolfe P 1982 Skin surface PO_2 measurements over the ischial tuberosity *Arch. Phys. Med. Rehabil.* **63** 553–6

Nicholson P W, Leeman A L, O'Neil C J A, Dobbs S M, Deshmukh A A, Denham M J, Royston J P and Dobbs R J 1988 Pressure sores: effect of Parkinson's disease and cognitive function on spontaneous movement in bed *Age and Ageing* **17** 111–5

Nimit K 1989 Guidelines for home air-fluidized bed therapy. *Health Technology Assessment Reports* (Rockville, MD: National center for health services research and health care technology assessment)

Noble P C 1981 *The Prevention of Pressure Sores in Persons with Spinal Cord Injuries* (New York: World Rehabilitation Fund)

Noble P C, Goode B L, Krouskop T A and Crisp B 1983 The response of polyurethane foam wheelchair cushions to environmental aging *Proc. Annu. Conf. Rehabil. Eng. Soc. N. Am.* **6** 218–20

Norton D, McLaren R and Exton-Smith A N 1962 *An investigation of geriatric nursing problems in hospitals* (Edinburgh: Churchill-Livingstone)

Nunn T A and Angell J B 1975 An IC absolute pressure transducer with built-in reference chamber *Indwelling and Implantable Pressure Transducers* ed D G Fleming, W H Ko, and M R Neuman (Cleveland, OH: CRC Press)

Oliver D E 1984 Pulsed electro-magnetic energy—what is it? *Physiotherapy* **70** 458–9

Olson W H 1978 Basic concepts of instrumentation *Medical Instrumentation Application and Design* ed J G Webster (Boston: Houghton Mifflin)

Palmieri V R, Haelen G T and Cochran G V 1980 A comparison of sitting pressures on wheelchair cushions as measured by air cell transducers and miniature electronic transducers *Bull. Prosth. Res.* **17** 5–8

Parish L C, Witkowski J A and Crissey J T (eds) 1983 *The Decubitus Ulcer* (New York: Masson Publishing)

Patel A, Kothari M, Webster J G, Tompkins W and Wertsch J J 1989 A capacitance pressure sensor using a phase-locked loop *J. Rehabil. Res. Dev.* **26** (2) 55–62

Patterson R 1990 Personal communication

Patterson R P 1983 Instrumentation for electrotherapy *Therapeutic Electricity and Ultraviolet Radiation* ed G K Stillwell (Baltimore: Williams & Wilkins)

Patterson R P 1984 Is pressure the most important parameter? *National symposium on the care, treatment and prevention of decubitus ulcers* 69–72

Patterson R P and Fisher S V 1979 The accuracy of electrical transducers for the measurement of pressure applied to the skin *IEEE Trans. Biomed. Eng.* **BME-26** 450–6

Patterson R P and Fisher S V 1980 Pressure and temperature patterns under the ischial tuberosities *Bull. Prosth. Res.* **17** 5–11

Patterson R P and Stradal L C 1973 Warning device for the prevention of ischaemic ulcers in patients who have injured the spinal cord *Med. Biol. Eng.* **11** 504–5

Paul B J, Lafratta C W, Dawson A R, Baab E and Bullock F 1960 Use of ultrasound in the treatment of pressure sores in patients with spinal cord injury *Arch. Phys. Med. Rehabil.* **41** 438–40

Pax R A Jr., Webster J G and Radwin R G 1989 A conductive polymer sensor for the measurement of palmar pressures *Proc. Annu. Int. Conf. IEEE Eng. Med. Biol. Soc.* **11** 1483–4

Pearce J A 1971 Skin pressure distributions on three methods of patient support *Air-fluidized Bed Clinical and Research Symposium* ed C P Artz and T S Hargest (Charleston, SC: Department of Surgery, Medical College of South Carolina)

Perkash I, O'Neill H, Politi-Meeks D and Beets C L 1984 Development and evaluation of a universal contoured cushion *Paraplegia* **22** 358-65

Pers M, Snorrason K and Nielsen I M 1986 Primary results following surgical treatment of pressure sores *Scand. J. Plast. Reconstr. Surg.* **20** 123–4

Petersen N C 1976 The development of pressure sores during hospitalisation *Bedsore Biomechanics* ed R M Kenedi, J M Cowden and J T Scales (Baltimore: University Park Press) 219–24

Petersen N C and Bittmann S 1971 The epidemiology of pressure sores *Scand. J. Plast. Reconstr. Surg.* **5** 62–6

Piche A, Rakheja S, Gouw G J and Sankar T S 1988 Development of an elastic human-seat interface pressure sensing system *Proc. Int. Conf. Am. Assoc. Rehabil. Therapy* 118–9

Pickett J M 1974 Research with the Upton eyeglass speechreader *Sensory aids for the hearing impaired* ed H Levitt, J M Pickett and R A Houde (New York: IEEE Press)

Pinchcofsky-Devin G D and Kaminski M V 1986 Correlation of pressure sores and nutritional status. *Am. Ger. Soc.* 34 435–40

Prockop D J, Kivirikko K I, Tuderman L and Guzman N A 1979 The biosynthesis of collagen and its disorders *New Eng. J. Med.* 301 13–23

Pye G and Bowker P 1976 Skin temperature as an indicator of stress in soft tissue *J. Inst. Mech. Eng.* 5 58–60

Reddy N P 1986 Mechanical stress and viability of skin and subcutaneous tissue *Tissue Nutrition and Viability* ed A R Hargens (New York: Springer-Verlag)

Reddy N P, Palmieri V and Cochran G V 1984 Evaluation of transducer performance for buttock-cushion interface pressure measurements *J. Rehabil. Res. Dev.* 21 43–50

Reddy N P, Patel H and Cochran G B 1982 Model experiments to study the stress distribution in a seated buttock *J. Biomech.* 15 493–504

Reger S I and Chung K C 1985 Comparative evaluation of pressure transducers for seating *Proc. Annu. Conf. Rehabil. Eng. Soc. N. Am.* 8 347–9

Reger S I, Chung K C, Martin G and McLaurin C A 1985 Shape and pressure distribution on wheelchair cushions. *Proc. Annu. Conf. Rehabil. Eng. Soc. N. Am.* 8 341–3

Reger S I, Chung K C and Pauling M 1986 Weightbearing tissue contour and deformation by magnetic resonance imaging. *Proc. Annu. Conf. Rehabil. Eng. Soc. N. Am.* 9 387–9

Regnault W F and Picciolo G L 1987 Review of medical biosensors and associated materials problems *J. Biomed. Mater. Res.: Appl. Biomaterials* 21 (A2) 163–80

Reichel S M 1958 Shear force as a factor in decubitus ulcers in paraplegics *JAMA* 166 762–3

Reichel S M 1958 Shearing forces as a factor in decubitus ulcers in paraplegics *JAMA* 166 172

Reichert H 1986 Surgical treatment of pressure sores in paraplegics and possible prevention of their recurrence *Scand. J. Plast. Reconst. Surg.* 20 125–7

Reswick J B and Rogers J E 1976 Experience at Rancho Los Amigos Hospital with devices and techniques to prevent pressure sores *Bedsore Biomechanics* ed R M Kenedi, J M Cowden and J T Scales (Baltimore: University Park Press)

Rhodes A, Sherk H, Black J and Margulies C 1988 High resolution analysis of ground foot reaction forces *Foot Ankle* 9 135–8

Roaf R 1976 The causation and prevention of sores *Bedsore Biomechanics* ed R M Kenedi, J M Cowden and J T Scales (Baltimore: University Park Press)

Robertson W V 1964 Metabolism of collagen in mammalian tissues. *Connective Tissue: Intercellular Macromolecules* (Boston: Little, Brown)

Robinson J 1981 *Understanding finite element stress analysis* (England: Robinson and Associates)

Rodriguez G P and Claus-Walker J 1988 Biochemical changes in skin composition in spinal cord injury: a possible contribution to decubitus ulcers *Paraplegia* 26 302–9

Rodriguez G P, Claus-Walker J, Kent M C and Stal S 1986 Adrenergic receptors in insensitive skin of spinal cord injured patients *Arch. Phys. Med. Rehabil.* 67 177–80

Roemer R B 1979 Oscillating wheelchair seat for prevention of decubitus ulcers *Med. Biol. Eng. Comput.* 17 379–82

Roemer R B, Meisel H and Parrish W J 1975 Automated bed to aid pulmonary drainage and prevent decubitus ulcers *Med. Biol. Eng. Comput.* 13 78–82

Rosen J M, Hentz V R and Perkash I 1986 The plastic surgical management of pressure sores *Proc. Annu. Conf. Eng. Biol. Med.* 28 281

Rosenberg L and Lach E 1985 Pressure sore prevention system: A new biofeedback approach [letter] *Plast. Reconstr. Surg.* 75 926–8

Ross C T F 1987 *Advanced applied stress analysis* (Chichester: Ellis Horwood)

Rothery F A 1989 Preliminary evaluation of a pressure clinic in a new spinal injuries unit *Paraplegia* 27 36–40

Rubin C F, Dietz R R and Abruzzese R S 1974 Auditing the decubitus ulcer problem *Am. J. Nurs.* 74 1820–1

Rudd T N 1962 Pathogenesis of decubitus ulcers *J. Am. Ger. Soc.* 10 48–53

Ryan D W 1989 A study of contact pressure points in specialised beds *Clin. Phys. Physiol. Meas.* 10 331–5

Ryan D W and Byrne P 1989 A study of contact pressure points in specialized beds *Clin. Phys. Physiol. Meas.* 10 331–5

Sabelman E E, Valainis E and Sacks A H 1989 Skin capillary blood flow under very light pressure *Proc. Annu. Conf. Rehab. Eng. Soc. N. Am.* 12 .256–7

Sacks A H 1989 Theoretical Prediction of a time-at-pressure curve for avoiding pressure sores *J. Rehab. Res. Dev.* 26 27–34

Sacks A H, O'Neill H and Perkash I 1985 Skin blood flow changes and tissue deformations produced by cylindrical indentors *J. Rehab. Res. Dev.* 22 1–6

Salmons S and Henriksson J 1981 The adaptive response of skeletal muscle to increased use *Muscle Nerve* 4 94–105

Scales J T 1976 Air support systems for the prevention of bed sores *Bedsore Biomechanics* ed R M Kenedi, J M Cowden and J T Scales (Baltimore: University Park Press)

Scales J T, Hopkins L A, Bloch M, Towers A G and Muir I F K 1967 Levitation in the treatment of large-area burns *Lancet* **10 June** 1235–40

Schulman D A 1989 *Fluid Pressurized Cushion* United States Patent 4,852,195

Sebern M D 1986 Pressure ulcer management in home health care: efficacy and cost effectiveness of moisture vapor permeable dressing *Arch. Phys. Med. Rehabil.* **67** 726–9

Seiler W O, Allen S and Stähelin H B 1986 Influence of the 30° laterally inclined position and the 'super-soft' 3-piece mattress on skin oxygen tension on areas of maximum pressure—implications for pressure sore prevention *Gerontology* **32** 158–66

Seiler W O and Stähelin H B 1979 Skin oxygen tension as a function of imposed skin pressure: Implications for decubitus ulcer formation *J. Am. Geriatr. Soc.* **27** 298–301

Seiler W O and Stähelin H B 1985 Decubitus ulcers: treatment through five therapeutic principles *Geriatrics* **40** 30–44

Seiler S O and Stähelin H B 1986 Recent findings on decubitus ulcer pathology: Implications for care. *Geriatrics* **41** 47–60

Sensym, Inc.1988 *1988 Sensym Pressure Sensor Handbook* Reamwood, CA

Seow K 1988 Capacitive Sensors *Tactile Sensors for Robotics and Medicine* ed J G Webster (New York: John Wiley & Sons)

Settle C M 1987 Sitting and pressure sores *Physiotherapy* **73** 455–7

Seymour R J and Lacefield W E 1985 Wheelchair cushion effect on pressure and skin temperature *Arch. Phys. Med. Rehabil.* **66** 103–8

Shield R K 1986 Effect of a lumbar on seated buttock pressure *Proc. Annu. Conf. Rehabil. Eng. Soc. N. Am.* **9** 408–10

Shield R K 1987 Effect of seat angle and lumbar support on seated buttock pressure *Proc. Annu. Conf. Rehabil. Eng. Soc. N. Am.* **10** 530–2

Shield R K and Cook T M 1988 Effect of seat angle and lumbar support on seated buttock pressure *Phys. Ther.* **68** 1682–6

Siedband M P and Holden J E 1978 Medical imaging systems *Medical Instrumentation: Application and Design* ed J G Webster (Boston: Houghton Mifflin)

Siegel R J, Vistnes L M and Laub D R 1973 Use of the water bed for prevention of pressure sores *Plast. Reconstr. Surg.* **51** 31–7

Sieggreen M Y 1987 Healing of physical wounds *Nurs. Clin. N. Am.* **22** 439–47

Silvetti A N 1981 An effective method of treating long-enduring wounds and ulcers by topical applications of solutions of nutrients *J. Dermatol. Surg. Oncol.* **7** 501–8

Smith M, Bowman L and Meindl J 1986 Analysis, design, and performance of a capacitive pressure sensor IC *IEEE Trans. Biomed. Eng.* **BME-33** 163–74

Snell L T 1989 Discrete motion on the anesthetized body in dorsolithotomy position as a means of diminishing intensity and/or occurrence of pressure-induced decubitus ulcers, Univ. Wis. Hosp. Clin., Madison WI 53706

Soames R W, Blake C D, Stott J R R, Goodbody A and Brewerton D A 1982 Measurement of pressure under the foot during function *Med. Biol. Eng. Comput.* **20** 489–95

Solis I, Krouskop T, Trainer N and Marburger R 1988 Supine interface pressure in children *Arch. Phys. Med. Rehabil.* **69** 524–6

Souther S G 1974 Wheelchair cushions to reduce pressure under bony prominences *Arch. Phys. Med. Rehabil.* **55** 460–4

Sprigle S and Chung K C 1989 The use of contoured foam to reduce seat interface pressures *Proc. Annu. Conf. Rehabil. Eng. Soc. N. Am.* **12** 242–3

Sprigle S, Chung K C and Brubaker C E 1990 Reduction of sitting pressures with custom contoured cushions *J. Rehabil. Res. Dev.* **27** 135–40

Sprigle S, Chung K C and Faisant T 1988 Clinical evaluation of custom contoured cushions on spinal cord injured subjects *Proc. Int. Conf. Am. Ass. Reh. Ther.* 264–5

Stapleton M F 1978 *Prevalence of Pressure Sores in One Health District* Worthing Health District

Stefanovska A, Vodovnik L and Kononenko I 1989 Toward a logic for the effect of electrical stimulation on wound healing *Proc. Annu. Conf. Rehabil. Eng. Soc. N. Am.* **12** 173–4

Stewart I M 1976 Sand bed nursing *Bedsore Biomechanics* ed R M Kenedi, J M Cowden and J T Scales (Baltimore: University Park Press)

Stone D R and Lambert C E 1975 *Orthopaedic Physician's Assistant Techniques* (Indianapolis: Howard W. Sams)

Stone N and Meister A 1962 Function of ascorbic acid in the conversion of proline to collagen hydroxyproline *Nature* **194** 555–7

Sugarman B 1985 Infection and pressure sores *Arch. Phys. Med. Rehabil.* **66** 177–9

Sutton R A, Minns R J, Sherrin A J, Brown J, Reignier J C and Herlant M 1987 Soft tissue augmentation as a method of reducing the liability to pressure sores in spinal injured patients *Paraplegia* **25** 454–65

Taylor E T, Rimmer S and Dias B 1974 Ascorbic acid supplementation in the treatment of pressure sores *Lancet* **2** 544–6

Taylor K J, Bryant R and Boarini J 1988 Assessment tools for the identification of patients at risk for the development of pressure sore: a review *J. Enterostomal Ther.* **15** 201–5

Tayton K, Philips G and Rallis Z 1982 Long term effects of carbon fiber on soft tissues *J. Bone Joint Surg.* **64B** 112–4

Thornhill-Joynes M, Gonzales F, Stewart C A, Kanel G C, Lee G C, Capen D A, Sapico F L, Canawati H N and Montgomerie J Z 1986 Osteomyelitis associated with pressure ulcers *Arch. Phys. Med. Rehabil.* **67** 314–8

Torrance C 1983 *Pressure Sores, Aetiology, Treatment and Prevention* (London: Croom Helm)

Trachtman L H, Ferguson-Pell M W, Palmieri V, Cardi M D and Cochran G V B 1984 Development of a computer-aided wheelchair cushion fitting clinic for the prevention of pressure sores *Proc. Int. Conf. Rehabil. Eng.* 463–4

Trandel R S 1975 Thermographical investigation of decubitus ulcers *Bull. Prosth. Res.* **24** 137–55

Traub W 1974 Some stereochemical implications of the molecular conformation of collagen *Israel J. Chem.* **12** 435–9

Turner R H, Kurban A K and Ryan T J 1969 Fibrinolytic activity in human skin following epidermal injury *J. Invest. Dermatol.* **53** 458–62

Turner T D 1982 Which dressing and why *Nursing Times* **78 (Wound Care No. 11 Suppl.)** 41–4

Upton H W 1968 Wearable eyeglass speech reading aid *Am. Ann. Deaf* **113** 222–9

Verhonick P J, Lewis D W and Goller H O 1972 Thermography in the study of decubitus ulcers *Nurs. Res.* **21** 233–7

Vinckx L, Boeckx W and Berghmans J 1990 Analysis of the pressure perturbation due to the introduction of a measuring probe under an elastic garment *Med. Biol. Eng. Comp.* **28** 133–8

Vodovnik L and Stefanovska A 1989 Restorative and regenerative functional electrical stimulation *Osaka Int. Workshop on FNS* Osaka, Japan 11–27

Wadsworth H and Chanmugam A P P 1983 *Electrophysical Agents in Physiotherapy* (Marrickville, Australia: Science Press)

Walker K A 1971 *Pressure Sores Prevention and Treatment* (London: Butterworth)

Waterlow J 1985 A risk assessment card *Nursing Times* **49** 49–55

Watkins D and Holloway A J 1978 An instrument to measure cutaneous blood flow using the Doppler shift of laser light *IEEE Trans. Biomed. Eng.* **BME-25** 28–33

Webster J G 1978 Measurement of flow and volume of blood *Medical instrumentation: application and design* ed J G Webster (Boston: Houghton Mifflin)

Webster J G 1989 A pressure mat for preventing pressure sores *Proc. Annu. Int. Conf. IEEE Eng. Med. Biol. Soc.* **11** 1485–6

Wertsch J J, Webster J G, Tompkins W J, Harris G F, Zhu H and Loftsgaarden J 1989 Development of a sensory substitution system for the insensate foot *J. Rehabil. Res. Dev., Rehabil. R&D Progress Reports* **26 Suppl.** 276

Wharton G W, Milani J C and Dean L S 1988a A comparative study of five low pressure mattresses *Paraplegia* **26** 112

Wharton G W, Milani J C and Dean L S 1988b Pressure sore profile: cost and management (abstract) *Paraplegia* **26** 124

Wheatley D W, Berme N and Ferguson-Pell M W 1980 A low-profile load transducer for monitoring movement during sleep *Exp. Mech.* 19–20

White G W, Mathews R M and Fawcett S B 1989 Reducing risk of pressure sores: effects of watch prompts and alarm avoidance on wheelchair push-ups *J. Appl. Behavior Analysis* **22** 287–95

Wilkins K E and Gibson D A 1976 The patterns of spinal deformity in duchenne muscular dystrophy *J. Bone Joint Surg.* **58-A** 24–32

Williams C M, Lines C M and McKay E C 1988 Iron and zinc status in multiple sclerosis patients with pressure sores *Eur. J. Clin. Nutr.* **42** 312–8

Williams R, Krouskop T, Noble P and Brown J 1983 The influence of inflation pressure on the effectiveness of air filled wheelchair cushions *Proc. Ann. Conf. Rehabil. Eng.* 215–7

Wilms-Kretschmer K and Majno G 1969 Ischaemia of the skin *Am. J. Pathol.* **54** 327–53

Witkowski J A and Parish L C 1982 Histopathology of the decubitus ulcer *J. Am. Acad. Dermatol.* **6** 1014–21

Woodbine A 1979 A survey in Macclesfield *Nursing Times* **75** 1128–32

Young J S and Burns P E 1981 Pressure sores and the spinal cord injured: part two *Model Systems' SCI Dig.* **3** 11–26, 48

Zacharkow D 1984 *Wheelchair Posture and Pressure Sores* (Springfield, IL: Charles C Thomas)

Index

Milton Keynes UK
Ingram Content Group UK Ltd.
UKHW040106071024
449327UK00019B/862